Non-Insect Pests and their Management

THE AUTHORS

Dr. S.B. Singh born in 1st July, 1963 in Rewa district of Madhya Pradesh. He passed his M. Sc, Ag. (Entomology) in 1984 from JN Krishi Vishwa Vidyalaya. Jabalpur and Ph.D. Degree in 1995 from Devi Ahilya Vishwa Vidyalaya, Indore. He was posted as research scholar in 1989 in AICRP on Honey Bee (Research and Training) in College of Agriculture, Indore of Madhya Pradesh. Presently he is posted as Assistant Professor (Entomology) in the department of entomology, College of Agriculture. Indore, M.P. He has gained wide experience of leaching, research and extension in his 27 years of services. He has published more than 20 research papers in repute of journals and a number of abstracts in many seminars and symposia. He has written seven manuals on different aspects of teaching. He has several extension articles and bulletin publication.

Dr. Shadanan Upadhyay born in 24th October, 1956. He passed his M.Sc. Ag. (Entomology) and Ph.D. Degree from J.N. Krishi Vishwa Vidyalaya, Jabalpur during 1979 and as in service candidate in 1991, respectively. He was posted as Junior Scientist in sugarcane research station, Jaora, district Ratlam of Madhya Pradesh. Presently he is posted as Professor and Head of section, Department of Entomology, College of Agriculture, Indore, M.P. He has more man 35 years experience of teaching, research and extension. He has published more than 35 research papers in repute of journals. He has also published several articles, bulletins etc.

Dr. R.K.Choudhary born in 16th July 1956 in Hoshangabad district of Madhya Pradesh He passed his M.Sc.Ag. Entomology in 1983 from J.N.Krishi Vishwa Vidyalaya Jabalpur and Ph.D.Degree in Life Science (Zoology) with specialization in Agriculture from Devi Ahilya Vishwa Vidyalaya, Indore. He joined in Indo-German Project, Poarkheda (Hoshangabad) as Research Associate (Entomology). Further was he posted as Junior Scientist at Operational Project, Guna. He worked in Tribal Agri. Res. Project on Millets at Dindori (Mandla), All India Coordinated Cotton Improvement Project, Khandwa, AICRP on Safflower, Indore, and AICRP on Sorghum (Millets) Indore. Presently he is working as Associate Professor Entomology. He worked on Soybean, Gram, wheat, Minor millets, Cotton, Safflower, Sorghum and some vegetables. He has gained vast experience in his 32 years services. He published more than 40 research papers in repute of journals and a number of abstracts in many seminars and symposia. He has written 4 manuals on different aspects of teaching, He has also several extension articles and bulletins.

Non-Insect Pests and their Management

Authors

Dr. S.B. Singh

Assistant Professor (Entomology)

Dr. Shadanan Upadhyay

Professor (Entomology)

Dr. R.K. Choudhary

Associate Professor (Entomology)

College of Agriculture, Indore
Rajmata Vijayaraje Scindia Krishi Vishwa Vidyalaya,
Gwalior, M.P., India

BIOTECH

2025

BIOTECH BOOKS®

ISBN 978-81-7622-402-4

Published by: **BIOTECH BOOKS®**
4762-63/23, Ansari Road, Darya Ganj,
New Delhi - 110 002
Phone: +91-011-23262132
E-mail: biotechbooks@yahoo.co.in

Typeset at: **Classic Computer Services**
Delhi - 110 035

Digitally Printed at: **Replika Press Pvt. Ltd.**

PRINTED IN INDIA

Preface

Since long period, the insect pests cause enormous loss to the agricultural crops. We concentrate towards the arthropods but little bit knowing about non insect pests. On this aspect there is a scanty literature and even scattered in different forms of information. The textbook entitled **"Non-Insect Pests and their Management"** has been written primarily for the students of under graduate of agriculture science. Apart from discussing the principles and methods of non insect pests management the biology, life cycle and behavior of individual pests have also been discussed and arranged in various chapters with main title and sub titles.

We hope that the students of under graduates of agriculture science and entomologists will find the book valuable and beneficial regarding information about non insect pests and their management. We are especially grateful to our colleagues at various institutions and universities for their help rendered for designing of the text. We owe our sincere thanks to the publisher for his whole hearted support.

Dr. S.B. Singh

Dr. Shadanan Upadhyay

Dr. R.K. Choudhary

Contents

Chapter 1

Mites and their Management

General Information

Mites, along with ticks, are small arthropods belonging to the subclass Acari (also known as Acarina) and the class Arachnida. The study of mites and ticks is called **acarology** (from Greek word *akari*, a type of mite; and *logia*).

Classification

Phylum: Orthopoda

Subphylum: Chelicerata

Class: Arachnida,

Sub class: Acari

Order: Acarina

Family: Acaridae, Tetranichidae and others

Acari

Acari (or **Acarina**) are a taxon of arachnids that contains mites and ticks. The diversity of the Acari is extraordinary and its fossil history goes back to at least theearly Devonian period. As a result, acarologists (the people who study mites and ticks) have proposed a complex set of taxonomic ranks to classify mites. In most modern treatments, the Acari is considered a subclass of Arachnida and is composed of two or three superorders or orders: Acariformes (or Actinotrichida),Parasitiformes (or Anactinotrichida), and Opilioacariformes; the latter is often considered a subgroup within the Parasitiformes.

Most acarines are minute to small (*e.g.*, 0.08–1.00 millimetre), but the largest Acari (some ticks and red velvet mites) may reach lengths of 10–20 millimetres. Over

50,000 species have been described and it is estimated that a million or more species may exist.

Diversity, Life Styles and Ecology

Mites are among the most diverse and successful of all the invertebrate groups. They have exploited an incredible array of habitats, and because of their small size (most are microscopic), go largely unnoticed. Many live freely in the soil or water, but there are also a large number of species that live as parasites on plants, animals, and some that feed on mold. It is estimated that 48,200 species of mites have been described. Mites occupy a wide range of ecological niches. For example, Oribatid mites are important decomposers and occur in many habitats. They eat a wide variety of material including living and dead plant and fungal material, lichens and carrion; some are even predatory, though no species of Oribatida mite are parasites.

Many mites who have been well studied are parasitic on plants and animals. One family of mites Pyroglyphidae, or nest mites, live primarily in the nests of birds and animals. These mites are largely parasitic and consume blood, skin and keratin. Dust mites, which feed mostly on dead skin and hair shed from humans instead of consuming them from the organism directly, evolved from these parasitic ancestors.

Insects may also be infested by parasitic mites. Examples are *Varroa destructor*, which attaches to the body of the honey bee, and *Acarapis woodi* (family Tarsonemidae), which lives in the tracheae of honey bees. There are hundreds of species of mites associated with other bee species, and most are poorly described and understood. They attach to the bees in a variety of ways. For example,*Trigona corvina* workers have been found with mites attached to the outer face of their hind tibiae.Some are thought to be parasites, while others beneficial symbionts. Mites also parasitize some ant species, such as *Eciton burchellii*.

Some of the plant pests include the so-called spider mites (family Tetranychidae), thread-footed mites (family Tarsonemidae), and the gall mites (family Eriophyidae). Among the species that attack animals are members of the sarcoptic mange mites (family Sarcoptidae), which burrow under the skin. Demodex mites (family Demodicidae) are parasites that live in or near the hair follicles of mammals, including humans. Acari are mites, except for the three families of ticks.

The tropical species *Archegozetes longisetosus* is one of the strongest animals in the world, relative to its mass (100 μg): It lifts up to 1,182 times its own weight, over five times more than would be expected of such a minute animal. Mites also hold the record speed; for its length, *Paratarsotomus macropalpis* is the fastest animal on Earth.

Acarines are extremely diverse. They live in practically every habitat, and include aquatic (freshwater and sea water) and terrestrial species. They outnumber other arthropods in the soil organic matter and detritus.

Many are parasitic, and they affect both vertebrates and invertebrates. Most parasitic forms are external parasites, while the free living forms are generally predatory and may even be used to control undesirable arthropods. Others are detritivores that help to break down forest litter and dead organic matter, such as skin cells. Others still are plant feeders and may damage crops.

The feather mites, *Acari astigmata,* are found on almost all species of birds, except for penguins, and are highly specialized for life on their host. They may feed on uropygial oil, skin flakes, fungus, bacteria, and feathers, depending on the taxon they belong to. Their lifestyle are affected by the microclimate (ambient temperature and relative humidity), like the seasonal change in temperature causes feather mites to shift their microhabitats on blue tits, however, there is no evidence on microclimate affecting mite diversity.

Medical Significance

The majority of mite species are harmless to humans, but a few species of mites can colonize humans directly, act as vectors for disease transmission, or cause or contribute to allergenic diseases.

Mites which colonize human skin are the cause of several types of skin itchy rashes, such as grain itch, grocer's itch, and scabies. *Sarcoptes scabiei* is a parasitic mite responsible for scabies which is one of the three most common skin disorders in children. Demodex mites, which are common cause of mange in dogs and other domesticated animals, have also been implicated in the human skin disease rosacea, although the mechanism by which demodex contributes to the disease is unclear.

Chiggers are known primarily for their itchy bite, but they can also spread disease in some limited circumstances, such as scrub typhus. The house-mouse mite is the only known vector of the disease rickettsial pox.

Dust mites cause several forms of allergic diseases, including hay fever, asthma and eczema, and are known to aggravateatopic dermatitis. House dust mites are usually found in warm and humid locations, including beds. It is thought that inhalation of mites during sleep exposes the human body to some antigens that eventually induce hypersensitivity reaction.

Difference between Mite and Insect

Sl.No.	Insect	Mite
1.	Body 3 segmented-Head, thorax, abdomen.	Body 2 segmented- Cephalothorax (Gnathosoma) and abdomen (Idiosoma)
2.	3 pairs of legs and 2 pairs of wings	4 Pairs of legs in abdomen only
3.	Antennae and compound eyes present	Absent, simple eyes present
4.	MP- Different types (Sub phylum-Mandibulata)	MP- Piercing and sucking type forcep type and called Chalcerae (Sub phylum-Chalicerata)
5.	Respiration by trachea	Respiration by book lungs
6.	5 segmented legs	2-8 segmented (Generally 7 segmented)

Morphology

Mites are arachnids and, as such, evolved from a segmented body with the segments organised into two tagmata: a prosoma (cephalothorax) and an opisthosoma (abdomen). However, only the faintest traces of primary segmentation remain in mites; the prosoma and opisthosoma are fused, and a region of flexible cuticle (the cirumcapitular furrow) separates the chelicerae and pedipalps from the rest of the body. This anterior body region is called the capitulum or gnathosoma and, according to some workers, is also found in Ricinulei. The remainder of the body is called the idiosoma and is unique to mites.

Most adult mites have four pairs of legs, like other arachnids, but some have fewer. For example, gall mites like*Phyllocoptes variabilis* (family Eriophyidae) have a worm-like body with only two pairs of legs; some parasitic mites have only one or three pairs of legs in the adult stage. Larval and prelarval stages have a maximum of three pairs of legs; adult mites with only three pairs of legs may be called 'larviform'. Also members of the Nematalycidae within Endeostigmata, who lives between sand grains, have often wormlike and elongated bodies with reduced legs.

The mouth parts of mites may be adapted for biting, stinging, sawing or sucking. They breathe through tracheae, stigmata (small openings of the skin), intestines and the skin itself. Species hunting for other mites have very acute senses, but many mites are eyeless. The central eyes of arachnids are always missing, or they are fused into a single eye. Thus, any eye number from none to five may occur.

Ontogeny

Acarine ontogeny typically consists of an egg, a prelarval stage (often absent), a larval stage (hexapod except in Eriophyoidea which have only two pairs of legs), and a series of nymphal stages. Any or all of these stages except the adult may be suppressed or occur only within the body of a previous stage. Larvae (and prelarvae) have a maximum of three pairs of legs (legs are often reduced to stubs or absent in prelarvae); legs are added at the first nymphal stage. Usually, a maximum of three nymphal stages are present and they are referred to in sequence as the protonymph, deutonymph, and tritonymph; however, some soft ticks have supernumerary nymphal stages. The females of some Tarsonemidae bear sexually mature young. If one or more nymphal stages are absent, then authors may disagree on which stages are present. Only the Oribatida pass through all developmental stages.

Economic Importance

Crop

Damage to crops is perhaps the most costly economic effect of mites, especially by the spider mites and their relatives (Tetranychoidea), earth mites (Penthaleidae), thread-footed mites (Tarsonemidae) and the gall and rust mites (Eriophyidae).

Human

Some parasitic forms affect humans and other mammals, causing damage by their feeding, and can even be vectors of diseases, such as scrub typhus, rickettsialpox, Lyme disease, Q fever, Colorado tick fever, tularemia, tick-borne relapsing fever, babesiosis, ehrlichiosis and tick-borne meningoencephalitis. A well known effect of mites on humans is their role as an allergen and the stimulation of asthma in people affected by respiratory disease.

Beneficial

The use of predatory mites (*e.g.*, Phytoseiidae) in pest control and herbivorous mites that infest weeds are also of importance. An unquantified, but major positive contribution of the Acari is their normal functioning in ecosystems, especially their roles in the decomposer subsystem.

Chemical agents used to control ticks and mites include dusting sulfur and ivermectin.

Based on habit and habitat mites are classified in following categories:

1. Phytophagous mites
2. Parasitic mites
3. Stored product mites
4. Soil mites.

1. Phytophagous Mites

Phytophagous mites are of agricultural importance as they cause economic damage to agricultural and horticultural crops which directly affects the human being. In modern agriculture the huge and imbalance application of pesticides and fertilizers, dense farming and poor management has increased the mite's presence besides other insect pests. Mites are wingless and due to that they are dispersed by human, other animals, birds blowing wind, mites infested plants and transportation of fruits and other materials.

Chauhan and Dahia (2001) reported following yield losses due to mites.

Crop	*Yield Loss (Per cent)*	*Crop*	*Yield Loss (Per cent)*
Mango	50-80	Litchi	30-40
Paddy	20-25	Pigeonpea	15-30
Sugarcane	20-30	Cotton	20-30
Tea	15-20	Ber	15-30
Vegetables	10-15		

Phytophagous mites are mainly divided into 4 families:

(i) Tetranichidae
(ii) Tenuipalpidae
(iii) Tarsonemidae
(iv) Eriophidae

(i) Tetranichidae

Spider mites are members of the Acari (mite) family **Tetranychidae,** which includes about 1,200 species. They generally live on the undersides of leaves of plants, where they may spin protective silk webs, and they can cause damage by puncturing the plant cells to feed. Spider mites are known to feed on several hundred species of plants.

Spider mites are less than 1 millimetre (0.04 in) in size and vary in color. They lay small, spherical, initially transparent eggs and many species spin silk webbing to help protect the colony from predators; they get the "spider" part of their common name from this webbing.

This family is a large group of plant feeding mites (about 1 200 species known worldwide), and can probably be regarded as the most important family in the Acari regarding members that are pests of agricultural crops and other plants.

The spider mites have round to oval-shaped bodies, and are as large as or smaller than a pin head (about 0.3 to 0.8 mm long). Their colour vary: green, red, brown, yellow, orange, black or combinations of any of these, depending on the species and the life stage.

Most tetranychid species have a wide host range, other than the generally more host specific Eriophyoidea. Some species, especially Tetranychus spp., can produce extensive webbing which impedes effective control of these pests. Chemical control of spider mites is also compounded by development of resistance to acaricides.

Life Cycle

Hot, dry conditions are often associated with population build-up of spider mites. Under optimal conditions (approximately 80 °F or 27 °C), the two-spotted spider mite can hatch in as little as 3 days, and become sexually mature in as little as 5 days. One female can lay up to 20 eggs per day and can live for 2 to 4 weeks, laying hundreds of eggs. This accelerated reproductive rate allows spider mite populations to adapt quickly to resist pesticides, so chemical control methods can become somewhat ineffectual when the same pesticide is used over a prolonged period.

Spider mites like hymenopterans and some scale insects, are arrhenotochous: females are diploid and males are haploid. When mated, females avoid the fecundation of some eggs to produce males. Fertilized eggs produce diploid females. Unmated, unfertilized females still lay eggs, that originate exclusively haploid males **(Plate 1).**

Plate 1: Spider Mites on a Lemon Plant.

Control Measures

- ☆ Neem oil may provide control, when combined with a suitable surfactant and diluted with water. As with chemical control, repeated applications are required.
- ☆ Predatory mites of the Phytoseiidae family, including *Phytoseiulus persimilis* eat adult mites, their eggs, and all developmental stages between. Predatory mites can consume as many as 5 adult spider mites per day, or 20 eggs per day.
- ☆ Insecticidal soap spray is effective against spider mites. It is commercially available or can be made of certain types of household soap. However, since it will also kill predatory mites, its use is not recommended if the latter are present.
- ☆ The mite's natural predator, *Phytoseiulus persimilis,* commonly used as a biological control method, is one of many predatory mites which prey mainly or exclusively on spider mites.
- ☆ Acaricides are applied to crops to control spider mites. They can be either systemic or non-systemic in nature and can be persistent by providing residual activity for over a month. Drawbacks include high potential for development of resistance in mite populations.

Red Spider Mite (*Tetranychus urticae*)

Introduction

Tetranychus urticae (common names include **red spider mite** and **two-spotted spider mite**) is a species of plant-feeding mite that is generally considered a pest. It is the most widely known member of the family Tetranychidae.

Identification

Tetranychus urticae is extremely small, barely visible with the naked eye as reddish or greenish spots on leaves and stems; the adult females measure about 0.4 mm long. The red spider mite, which can be seen in greenhouses and tropical and temperate zones, spins a fine web on and under leaves. These little red devils are tiny, crawling, wingless, insects that have 6 or 8 legs depending on their stage of development, and measure about 0.5 mm in length. In fact they are so tiny, need a ten power glass just to see them. Their presence is indicated by a very fine light speckling, or as localized pale yellow spots on the upper surfaces of leaves or in the case of cacti, plant bodies. Extensive damage is caused to leaves, buds and flowers with leaves eventually becoming bronzed, disfiguring plants for years to come and in severe cases causing plants to whither and die. These little red devils really do a number on plants. Once they sink their teeth into plant, the plant seems to go immediately into dormancy while they continue to eat away, and plants will only wake up and begin growing again once eradicate the problem.

Colour

Red spider mite is only red at certain stages of its life, when entering hibernation and in the springtime. A newly hatched mite is almost white, while the adult can be almost white in varying degrees through to green, with two spots on its back. These spots are said to be very dark green or dark red. The image here below must have been maginified a million time.

Damage

Tetranychus Urticae is extremely polyphagous. It can feed on hundreds of plants. These include most vegetables and food crops including peppers, tomatoes,potatoes, beans, maize and strawberries and ornamental plants such as roses. It is the most prevalent pest of *Withania somnifera* in India. It lays its eggs on the leaves, and it poses a threat to host plants by sucking cell contents from the leaves cell by cell, leaving tiny pale spots or scars where the green epidermal cells have been destroyed. Although the individual lesions are very small, attack by hundreds or thousands of spider mites can cause thousands of lesions and thus can significantly reduce the photosynthetic capability of plants.As with most pests they are often found on the underside of leaves, but on cacti they tend to encompass the entire body. On close inspection of leaves, may see the effects of the eggs from above, since the young hatchlings eat away at the underside of the leaf, sucking the life out of the plant while leaving yellowish spot clusters showing through. On cacti though, these unsightly clusters of pin prick sized spots begin showing themselves all over the plant body, eventually matting and looking like a great big tan coloured scar that can greatly reduce the photosynthetic capability of plants, as well as the production of nutrients and possibly even killing it. As adults, spider mites colonise the plant and may also spin a very fine web over the leaf or plant and eventually to other plants in collection. This webbing tends to act as a shelter for the young considerably reducing the effect of insecticides. Without lens these little red devils can be identified with a piece of bright white paper under direct sunlight, or a strong lamp by shaking the suspected leaves or plant itself over the paper and watching for very tiny specks scurrying to get to the other side. Mites are programmed to always go to the underside of a leaf if possible. Mite damage appear on the lower, older sections of plants and leaves first.

Life Cycle

Tetranychus urticae reproduces through arrhenotoky, a form of parthenogenesis in which unfertilized eggs develop into males. The eggs of *T. urticae* are translucent and pearl-like. It hatches into a larva and then two nymph stages follow: a protonymph, and then a deutonymph, which may display quiescent stages. The adults are typically pale green in colour for most of the year, but later generations are red in colour; mated females survive the winter in diapause. All spider mites go through the same stages of development. Adult females lay eggs on their host plants. The eggs hatch within days to weeks into the first stage called a larva with only three pairs of legs. The larva feed, then seek a sheltered spot to rest and moult into the first nymph stage having four pairs of legs. The nymphs then feed, rest and moult once again into the second nymph stage, feeding and resting before moulting into the adult stage. Males are smaller than females and have pointed abdomens, while females have rounded abdomens and are the largest mites present.

Realistically, heat speeds up the life cycle of these dreaded parasites. It can be as short as three days in hot places, and as long as a month in cool weather with a dry atmosphere. As greenhouses tend warm up in the spring the breeding cycle begins, and will continue until autumn when plants begin to enter their dormancy cycle. The remaining females then turn red and find places to hibernate in locations such as the soil or compost of the host plant, the bark of mature plants and even in the wood or brickwork of the building itself. They are so tiny that no crevice is too small. Then in the spring they show themselves once again as bright orangey red devils attacking plants. As mentioned, hot, dry conditions are often associated with an acceleration of spider mite infestations. Under optimal conditions of approximately 80°F (25°C), they can hatch in as little as 3 days and become sexually mature in as little as five days. Under such conditions, a single female can lay up to 20 eggs per day and can live for two to four weeks, laying hundreds of eggs that in the long run can create a population explosion within just one month of over a million little devils eating away at plants. This accelerated reproductive rate allows spider mite populations to adapt quickly and build immunities to pesticide use over time. Cooler temperatures

Plates 2 and 3: Spider Mites.

Plates 4 and 5: Spider Mites.

and a moist atmosphere is known to slow down their metabolism and slow the reproductive rate, so regular misting may help with control.

Seasonal History

Mites come from the pits. Seriously in short, theses little red devils hitchhike on birds, furry animals, people (mostly in the hair and clothing) and insects including that of white fly, aphids, wasps and big bumble bees. Insects bring more disease and infestation into household than any other method, and big bees can do considerable damage to carefully tended show plants in a very short time with such devilish introductions **(Plates 2-5)**.

Carmine Spider Mites, *Tetranichus cinnabarinus* (Boisduval)

Symptoms

T. cinnabarinus often seems to cause damage at lower population densities than *T. urticae*. On tomatoes, typical spider mite speckling symptoms do not appear but, instead, infested leaflets become prematurely chlorotic with small transparent lesions. The damage closely resembles magnesium deficiency. Bright-yellow patches develop on the leaflets and necrotic patches from these gradually coalesce until the whole leaf withers and dies. The damage seems to be caused by a toxin injected in the saliva. On cotton, leaves may become discoloured and fall prematurely.

Integrated Pest Management

Prevention

- ☆ Prevention is better than a cure, so place "bug screens" over open windows and doors to stop pests from entering.
- ☆ Remove any infected plants immediately and burn it alive so the infestation doesn't spread.
- ☆ Check plants regularly for any signs of infestation. Do not buy infected plants. And do not bring infected plants to any plant sales or shows.
- ☆ If spider mites are found, try washing them off with a strong water sprayer. Mites, like aphids are easily dislodged from the leaf surface, at least before they have a chance to begin spinning their web. A regular hard blast of water once or twice a week can stop an infestation from occurring or slow it down once it starts, spray the undersides of the leaves.
- ☆ Improve the conditions in greenhouse with misting and cooler temperatures if possible.

Biological Control

- ☆ The predatory phytoseiid mite ***Phytoseiulus persimilis*** has been used with success in greenhouses to control *T. cinnabarinus*, in pure populations and when mixed with *T. urticae*, on a number of crops. This predator proved effective on cucumbers and on sweet pepper (*Capsicum annuum*). The adult phytoseiulus are slightly larger than red spider mites with longer legs and

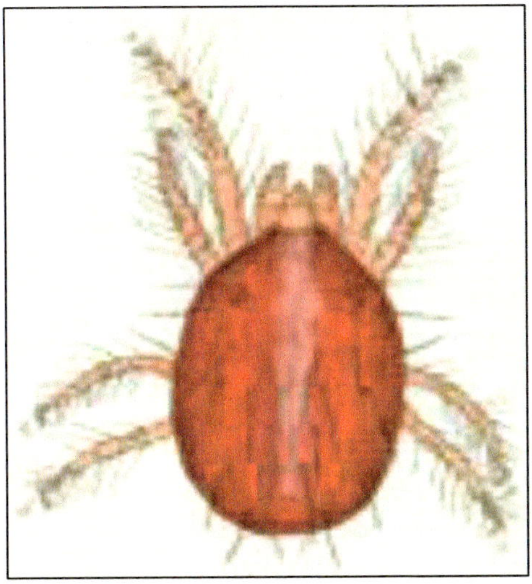

Plate 6: Adult.

Plate 7: Eggs.

Plate 8 Nymph and Adult.

Plate 9: Damage on Leaf.

Plates 6-9: Carmine Spider Mites.

an orangey red body. They move rapidly over plants, searching for their prey and each individual can consume 5 adults or 20 immature stages of red spider mite per day. Predatory mites eat adult mites, their eggs, pupae and all developmental stages. Other species of predatory phytoseiid mites have also been tried. *Amblyseius gossypii* was released against *T. cinnabarinus* and *T. urticae* on cotton in, and an early release at the highest rate gave good control showed that *Typhlodromus athiasae* could suppress *T. cinnabarinus* on lemon seedlings and Anthoseius recki was effective on strawberry.

- ✰ Spider mites have many natural enemies such as mantis, dragonfly, wasp, spider, caterpillar, ladybugs

Cultural Control

- **Host-PlantResistance:** Differences in resistance to *T. cinnabarinus* have been shown for selections of melon, aubergine and cotton and cultivars of strawberry and taro. Kielkiewicz (1994) found that trichomes on the leaves of tomato contributed to a reduction in the density of mite.
- High levels of nitrogen in the leaves can increase spider mite reproduction rates.

Chemical Control

- Because of the problems with resistance to acaricides it is vital to use IPM programmes wherever possible, to maximize the regulating effect of natural enemies.
- Using insecticides, and insecticidal soaps to control Red Spider Mites.Adding a little insecticidal soap and methylated spirits to the water in mister will also aid in the battle against this most persistent pest.
- Remove the webs and the spider mites by washing the plant in soapy tepid water, paying particular attention to the underside of all the leaves. Repeat the treatment every 3 to 4 days to ensure its success. Washing will not always remove all the eggs so repeat on a weekly basis until the plant is clear. For severe infestations it is best to destroy the plant.
- Use miticides for heavy infestation that are specifically used to control red spider mites. Insecticides against normal bugs are ineffective in killing red spider mites, and may aggravate the problem as the insects that eat the mites are killed by the chemicals instead.

  Ethion 50EC @ 1000ml/ha.

  Dicofol 18.5EC @1500ml/ha

  Propargite 57EC @ 750-1000ml/ha

  Dimethoate 30EC @ 1000ml/ha

  Apply neem products @ 5ml/L on vegetables.
- Use of mythel hydrate, sprayed directly on the plant which kills red spider mites on contact without harming plants. Repeat the treatment every three to four days until the situation is under control.
- Hot pepper wax, which is sprayed to prevent transpiration is another remedy which smothers the adults on the leaves, but does not kill the eggs so be prepared to repeat applications every four to five days until control is established. Cayenne powder, sprayed onto the surface of the compost in your pots, is also reputed to deter crawling insects and sciarid flies.

(ii) Tenuipalpidae

Tenuipalpidae, also called **flat mites** or **false spider mites**, are a family of mites, closely related to the Tetranychidae. They are reddish and slow-moving and normally feed near the midrib or veins on the underside of leaves.

The false spider mites are very small (females about 0.2 - 0.4 mm long) (smaller than the spider mites) and somewhat flattened, and are difficult to see with the naked eye. They are usually reddish or brown red. They move relatively slowly and lie very flat against the plant surfaces. Tenuipalpid mites appear to be best adapted to subtropical or tropical regions. Members of the genera Brevipalpus, Tenuipalpus and Dolichotetranychus are of particular importance as plant pests. In southern Africa, four species are regarded as pests of economic importance:

(iii) Tarsonemidae

Floriculture and Ornamental Nurseries Thread-footed (Tarsonemid) Mites

Cyclamen mite: ***Phytonemus pallidus***

Broad mite: ***Polyphagotarsonemus latus***

Bulb scale mite: ***Stenotarsonemus laticeps***

Description of the Pests

Thread-footed mites are tiny and cannot be readily seen without magnification (20X to 40X). The life stages of these thread-footed mites are: egg, nymph, pseudopupa, and adult (one less stage than for spider mites). Eggs of the cyclamen mite are one-half the length of the adult and are oval shaped. Eggs of the closely related broad mite are distinguishable from cyclamen mite eggs by rows of white pegs on the egg's upper surface. Immature stages of these mites are white.

These mites get the name "thread-footed" from the appearance of the hind pair of legs in the adult female, which is threadlike; adult males, on the other hand, have stout legs for clasping the female. Adult males carry female pseudo pupae on their back. As soon as the adult female emerges, they mate.

Cyclamen mite is generally found feeding on growing terminals, in buds, or on unfolding leaflets. Their development is optimal under moderately warm (60° to 80°F) temperatures and high humidity (80 to 90 per cent). Broad mite is similar to cyclamen mite, but is generally found mostly on the undersides of plant leaves. Mites disperse between plants on air currents and by mechanical transport such as on worker's clothing. These mites can complete one generation in 7 to 21 days, depending on temperature. Female bulb scale mites lay up to 28 eggs. Adults are usually found between the scales of the bulb and the neck region. Bulb mites overwinter in bulbs between the scales, emerging as the leaves grow. They re-enter bulbs as the bulbs dry in the field. One generation can be completed in about 7 weeks under field conditions.

Damage

Feeding by cyclamen and broad mites is easily recognized on all hosts because affected leaves become characteristically cupped, dwarfed and thickened, and the internodes are greatly shortened. Broad mite damage occurs more generally over the plant than cyclamen mite damage. Bulb scale mites feeding in developing shoots can cause longitudinal bronze streaks of discoloration, horizontal cracks, distortion, and death of leaves and flowers **(Plates 10-11).**

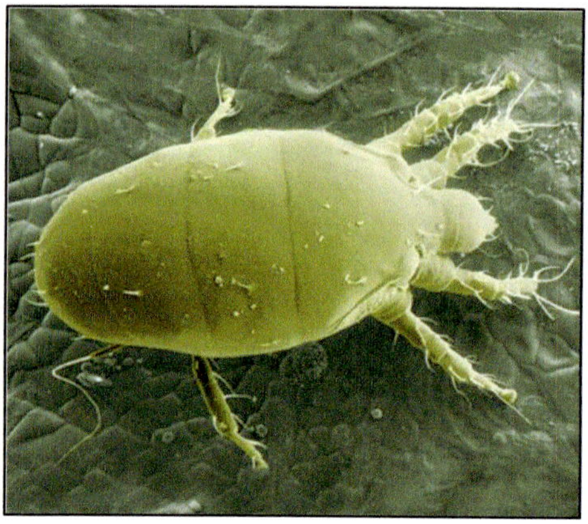

Plate 10: Tarsonemid Mite Adult.

Plate 11: Damage by Tarsonemid Mite.

Management

Biological Control

- *Neoseiulus californicus* and other species of predatory mites have been used for broad mite and cyclamen mite control.
- Three entomogenous fungi, *Beauveria bassiana*, *Metarhizium anisopliae*, and *Paecilomyces fumosoroseus*, can effectively manage broad mite infestations, with *B. bassiana* providing the greatest reduction. Predatory phytoseiid mites, in the genus *Neoseiulus*, can also successfully control *P. latus* and *S. pallidus* under greenhouse and field conditions.

Cultural Control

Because these mites feed on a large variety of plant species, keep production areas free of weeds that can serve as hosts for mite populations. Carefully inspect plants being brought in to start a new crop to ensure that they are free of pests, and disinfest the plants if needed. Disinfestation can be accomplished by immersing propagation stock in 110.3°F water for 30 minutes, or treatment at 100 per cent relative humidity and 110.3°F for 1 hour. If hot spots of these mites are found in production areas, consider roguing affected plants and treating the surrounding plants.

Chemical Control

- **Abamectin (Avid 0.15EC) (Syngenta)- macrocyclic lactone**- low-volume application.
- **Clarified hydrophobic extract of neem oil (Triact 70**)- Do not spray plants under stress. Target pest must be completely covered with spray.
- Chemical trials demonstrated that endosulfan and dicofol consistently reduced densities of *P. latus* and *S. pallidus*, and planting stock can be effectively decontaminated through fumigation with methyl bromide or 1,2-dibromoethane.

Tarsonemidae is a family of mites, also called **thread-footed mites** or **white mites**. Only a limited number of tarsonemid genera (*Steneotarsonemus, Polyphagotarsonemus, Phytonemus, Floridotarsonemus* and *Tarsonemus*) are known to feed on higher plants while most species in this family feed on the thin-walled mycelia of fungi or possibly algal bodies. Even among the plant-feeding tarsonemid mites, most are confined to areas of new growth where cell walls are thin and therefore easily pierced. However two species (the "broad mite" *Polyphagotarsonemus latus* and the "cyclamen mite" *Steneotarsonemus pallidus*) are able to feed on older leaves because of their ability to inject toxins during feeding (presumably of salivary gland origin) causing an increase of thin walled cells surrounding feeding sites. This proliferation of new growth often results in leaves that appear stunted, puckered and twisted.

Selected Genera

Acarapis

Floridotarsonemus

Phytonemus

Polyphagotarsonemus

Steneotarsonemus

Tarsonemus

(iv) Eriophidae

Plant Feeding Mites

More than 6 000 species of plant feeding (phytophagous) mites are known worldwide. The majority of plant feeding species belong to the obligate plant parasitic Eriophyoidea (*e.g.* gall mites, erinose mites, bud mites, rust mites) and Tetranychoidea (*e.g.* spider mites, false spider mites), while a number of species belong to other lineages (*e.g.* Eupodoidea, Tarsonemidae, and single oribatid mites) **(Plates 12-16).**

Plant feeding mites form an integral and important part of the natural ecosystem. Some species, especially eriophyoid mites, can be utilised for the biological control of weeds. Many plant feeding mites are of economic importance as pests of crop plants.

- ☆ All eriophyoid mites are plant feeding. They are extremely tiny, the majority are less than 300 micron long, and essentially invisible to the unaided eye. They have elongated, worm-like bodies, with only two pairs of legs.
- ☆ Eriophyoid mites are commonly known as gall mites, bud mites, rust mites, erineum mites, witches' broom mites, blister mites and so on, referring to the symptoms caused by a particular species. The feeding of almost half of eriophyoid species known, though, does not cause visible damage to their plant hosts.
- ☆ The eriophyoid mites belong to three families: Phytoptidae, Eriophyidae and Diptilomiopidae. About 3 400 species are known, but these probably represent only as little as 5 per cent or less of extant eriophyoid species. Most woody and many herbaceous flowering plants, and gymnosperms and ferns most likely host these minute mites. Most eriophyoid species are highly host specific, restricted to one or single closely related plant species.

Plates 12 and 13: Damage by *Aceria saccha.*

Plate 14: Adult *Aceria sacchari.*

Plate 16: Damage by *Aceria sorghi* on Bajra.

Plate 15: Damage by *Aceria sorghi* on Sorghum.

Images of *Aceria sacchari*-Sugarcane Mite

Store Grain Mite, *Acarus siro* (L)

Classification

Secondary pest; Grain and fungal feeder

Order: Acarina

Family: Acaridae

Description (Plates 17-22)

- ☆ Mites are white and very small (less than 1 mm in length).
- ☆ Mites are soft bodied and have only a single body part.
- ☆ Nymphs have 6 legs and adults have 8 legs.

Commodities Affected

- ☆ Flour, grain, seeds and bulbs
- ☆ Also fungi that develop on these products

Plates 17 and 18: Adult–*Acarus siro.*

Plate 19: Nymph–*Acarus siro.*

Plate 20: Deutonymph of *Acarus siro.*

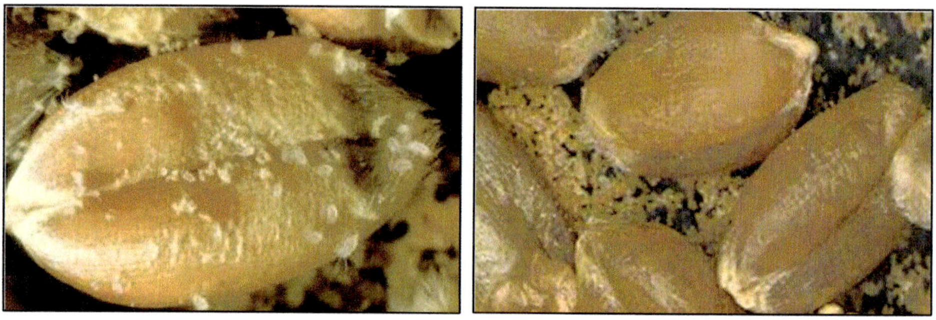

Plates 21 and 22: Damage by *Acarus siro.*

Signs of Infestation

- ☆ Grain dust may appear to be moving, actually mites are moving.
- ☆ Very high mite populations cause a sweet odour.

Damage

- ☆ Germ of grain is often completely destroyed
- ☆ Food may be rendered unusable for consumption

Geographic Range

- ☆ Is found worldwide and across Canada

Life History

- ☆ Females may lay up to 250 eggs in a month.
- ☆ Some mites may also enter a highly resistant, non-feeding resting stage known as a hypopus.
- ☆ Mites thrive in high humidity conditions.
- ☆ Mites mainly feed on the germ.
- ☆ Mites can subsist on organic debris in cracks, crevices and corners of storage areas.

Management

Residual Insecticides for structures

There are several insecticides available for treating cracks and crevices in buildings or empty bins and other storage containers:

- ☆ Malathion
- ☆ Pyrethrins
- ☆ Cyfluthrin
- ☆ Diatomaceous earth
- ☆ Spraying a chemical insecticide on the interior surface of a steel grain bin

Residual Insecticides for Direct Application to Grain

Diatomaceous Earth

- ☆ Absorbs waxy layer on insect skin causing death by desiccation of insect.
- ☆ Most effective when applied as dry grain is augered into a bin during harvest.
- ☆ Treatment for a 6-week period at above 20°C is necessary to ensure maximum efficacy.
- ☆ Diatomaceous earth has the potential to reduce test weight. Lower auger flow rates when applying as diatomaceous earth increases grain friction.

Malathion

- ☆ Incapacitates the insect's central nervous system.
- ☆ Best applied as a liquid spray or dust as grain flows through an auger or conveyer.
- ☆ Intended for use on cereal grains. It is not registered for use on oilseeds.
- ☆ Ensure that the formulation used is registered for direct grain treatment.
- ☆ Not all formulations of malathion meet this registration requirement.
- ☆ Grain treated with malathion should not be sold for 7 days, nor used for feed until 60 days after application.
- ☆ There is widespread resistance in insect populations to malathion.

Aluminum Phosphide (Phosphine)

- ☆ Applied as pellets to bulk grain either as the grain is entering storage; or placed into the grain with a probe after the grain is stored; or contained in prepackaged devices (ropes) on grain surface.
- ☆ Structures must be well sealed in order for treatment to be effective.
- ☆ Aluminum phosphide reacts with water in the air to produce the gas phosphine.
- ☆ Grain temperature and moisture are both factors in contributing to the effectiveness of aluminum phosphide. Grain should not be fumigated if the temperature is less than 5oC or if the moisture content is less than 10 per cent.
- ☆ Eggs are the most resistant stage, requiring approximately 20 times higher doses to control than adults and larvae.
- ☆ Phosphine used according to the label will control all life stages and all stored-grain insect pests.
- ☆ Monitoring of phosphine during fumigation will ensure that gas levels are sufficient to control insects.
- ☆ Phosphine may cause corrosion of certain metals (copper, brass, silver and gold) at high temperatures and humidity. Precautions should be taken to remove or protect equipment containing these metals such as electrical motors, wiring and electronic systems.

- ☆ Insect resistance to phosphine is known to occur in the United States and Australia and continuous or inappropriate use may result in resistance developing.

Magnesium Phosphide (phosphine)

- ☆ Like aluminum phosphide, magnesium phosphide reacts with water in the air to produce the gas phosphine.
- ☆ Magnesium phosphide generates phosphine faster that aluminum phosphide.
- ☆ Unlike aluminum phosphide, magnesium phosphide cannot come in contact with the commodities fumigated.
- ☆ Magnesium phosphide is applied contained in plates, pouches or strips.
- ☆ Mainly used for fumigating empty structures or tarped commodities.
- ☆ Similar issues with temperature, humidity and corrosion of metals as aluminum phosphide.
- ☆ Structures must be well-sealed in order for treatment to be effective.

Gaseous Phosphine

- ☆ Gaseous phosphine (compressed gas in a cylinder) allows a more precise dosage of phosphine and faster fumigation than the metal phosphides.
- ☆ Fumigation at minimum temperature of 0°C, compared to metal phosphides with a minimum temperature of 5°C.
- ☆ Similar issues with corrosion of metals as aluminum phosphide.
- ☆ 2 formulations: ECO2FUME® (ready for application), VAPORPH3OS® (needs onsite dilution).

Carbon Dioxide

- ☆ Carbon dioxide is applied to grain as a gas from cylinders
- ☆ Structures must be well sealed in order for treatment to be effective
- ☆ Dose: Maintain at 60 per cent for 4 days between 20°C and 25°C.

Methyl Bromide

- ☆ Methyl bromide can be used for quarantine and pre-shipment applications either for export or for import.
- ☆ The only quarantine insect for stored grain is the khapra beetle (*Trogoderma granarium*).There are several other non-grain pests that are found on wood or other commodities that may need to be fumigated before export or import into Canada.
- ☆ In the past, methyl bromide was used to fumigate flour mills and structures and grain to control stored grain insect pests, but this use has be almost entirely stopped because of restrictions under the *Montreal Protocol on Substances that Deplete the Ozone Layer*.

Physical Control of Grain Insect Pests

High Temperature

In the past, grain dryers have been used to help with disinfestation. There are a number of variables that are difficult to manage and may affect end use quality of grain, including temperature, time, and type of dryer. As a result, insect control using high temperature may produce variable results.

Caution: After you apply heat treatment, grain must be immediately cooled so that it does not overheat, spoil and cause a new insect infestation. Energy costs for this method of control are high. Therefore, using a grain dryer is recommended only when you need to rapidly dry grain.

Low Temperature

Producers and grain handlers can use ambient cold air to prevent and control infestations.

Stored product insect pests generally do not feed or reproduce at temperatures below 18°C. Lower temperatures can also be used to cause mortality. For example, grain kept at -5°C for 12 weeks will control stored insect pests at all life stages.

Lower the temperature of stored grain by:

- ☆ Mixing and transferring infested grain from one bin or pile to another
- ☆ Transferring part of the crop to a truck or small pile to expose it to low air temperature and leaving it to cool before returning it to the bin
- ☆ Aerating the bin

Aeration systems are very effective at reducing grain temperatures, as well as reducing moisture migration.

Time Required at Low Temperatures to Control Stored Grain Insect Pest Infestations

Constant Grain Temperature (°C)	*Time Period Required for Disinfestations (Weeks)*
-5	12
-10	8
-15	4
-20	1

Chapter 2
Plant Parasitic Nematodes

1. Root Knot Nematode

Meloidogyne incognita (Kofoid and White, 1919) Chitwood 1949

Taxonomic Tree

Domain: Eukaryota

Kingdom: Metazoa

Phylum: Nematoda

Family: Meloidogynidae

Genus: Meloidogyne

Species: *Meloidogyne incognita*

Introduction

Root-knot nematodes are plant-parasitic nematodes from the genus *Meloidogyne*. They exist in soil in areas with hot climates or short winters. About 2000 plants worldwide are susceptible to infection by root-knot nematodes and they cause approximately 5 per cent of global crop loss. Root-knot nematode larvae infect plant roots, causing the development of root-knot galls that drain the plant's photosynthate and nutrients. Infection of young plants may be lethal, while infection of mature plants causes decreased yield.

The root knot nematode species, *M. incognita*, is the most widespread and probably the most serious plant parasitic nematode pest of tropical and subtropical regions throughout the world. It occurs as a pest on a very wide range of crops. Most estimates of yield loss come from the use of nematicides and it should be noted that these can possibly cause other beneficial growth effects.

Meloidogyne incognita is a nematode (roundworm) in the family Heteroderidae. It is commonly called the "southern root-knot nematode" or the "cotton root-knot nematode". This parasitic roundworm has worldwide distribution and numerous hosts. It is an important plant parasite classified in parasitology as a root-knot nematode, as it prefers to attack the root of its host plant. When *M. incognita* attacks the roots of plants, it sets up a feeding location, where it deforms the normal root cells and establishes giant cells. The roots become gnarled or nodulated, forming galls, hence the term "root-knot" nematode.

M. incognita has been found to be able to move along shallower temperature gradients (0.001C/cm) than any other known organism, an example of thermotaxis. The response is complicated and thought to allow the nematodes to move toward an appropriate level in soil, while they search for chemical cues that can guide them to specific roots.

Economic Impact

Root-knot nematodes (*Meloidogyne* spp.) are one of the three most economically damaging genera of plant-parasitic nematodes on horticultural and field crops. Root-knot nematodes are distributed worldwide, and are obligate parasites of the roots of thousands of plant species, including monocotyledonous and dicotyledonous, herbaceous and woody plants. The genus includes more than 90 species, with some species having several races. Four *Meloidogyne* species (*M. javanica, M. arenaria, M. incognita, and M. hapla*) are major pests worldwide, with another seven being important on a local basis. *Meloidogyne* occurs in 23 of 43 crops listed as having plant-parasitic nematodes of major importance, ranging from field crops, through pasture and grasses, to horticultural, ornamental and vegetable crops. If root-knot nematodes become established in deep-rooted, perennial crops, control is difficult and options are limited.

Control

- ☆ Root-knot nematodes can be controlled with biocontrol agents *Paecilomyces lilacinus, Pasteuria penetrans* and Juglone.

Distribution

M. incognita has also been recorded from the following areas of protected agriculture; for example, around temperate research stations and other areas of intensive horticulture. In some cases the nematodes have been able to overwinter due to mild winters and could be expanding their geographical range.

Description

Female

Endoparasitic,body spherical with projecting neck. Head with 2 or occasionally 3 annules behind head-cap. Cuticle thickening abruptly at base of relaxed stylet. Stylet knobs rounded or drawn out laterally. Excretory pore at level of or posterior to stylet knobs, 10-20 annules behind head. Posterior cuticular pattern displays bewildering variation. Typical pattern 'incognita type' with striae closely spaced, very wavy to zig-zag especially dorsally and laterally. Dorsal arch high, rounded. Lateral field not clear, sometimes marked by breaks in striae, broken ends often forked, pattern merging into body striae. 'Acrita type' with striae smoother, more widely spaced (or with coarse widely spaced striae separated by fine closely spaced striae visible for short distances). Dorsal arch variable, may be flattened at top or trapezoid. Striae often forked along a 'lateral line'.

Male

Head not offset, a high truncate cone shape, clearly annulated. Head-cap with stepped outline in lateral view; annule number behind head-cap very variable, usually 1-3 on sublateral head sectors and 1-5 on lateral head sectors. Conus of stylet longer than shaft, stylet knobs prominent, usually of greater width than length, with flat, concave or 'toothed' anterior margins. Excretory pore at level of posterior end of isthmus with hemizonid usually 0-5 annules anterior. Lateral field with 4 incisures, outer bands areolated, inner band rarely cross-striated except at posterior end. Testes 1 or 2. Tail bluntly rounded, terminus unstriated. Phasmids at cloaca level or just anterior. Spicules slightly curved, gubernaculums crescentic.

Juvenile

Head not offset, truncate cone shape in lateral view, sub-hemispherical in dorso-ventral view. Head-cap wide followed by 2 clear annules on sublateral head sectors (lateral view), 3 annules on lateral head sectors (dorsal or ventral view). Stylet knobs prominent, rounded. Hemizonid 3 annules long just anterior to excretory pore. Lateral field with 4 incisures, outer bands cross striated. Rectum inflated. Tail tapering to subacute terminus, striae coarsening posteriorly.

Biology and Ecology

The life cycles of Meloidogyne spp. are well studied and in their essentials differ little between the major species. At 21°C *M. incognita* took 37 days to complete its life cycle on *Antirrhinum majus*, a similar time to that reported on soyabeans (, Juveniles penetrate root tips, occasionally invading roots in the zone of root elongation. Invaded nematodes initiate the development of giant cells in the meristematic, cortical and xylem tissues of the root and galling of roots occurs. Third- and fourth-stage juveniles and young females occur after about 6-8 and 15 days, respectively. Adult females were observed after 20 days and egg laying commenced after 25 days **(Plates 23-24).**

Life Cycle

All nematodes pass through an embryonic stage, four juvenile stages (J1–J4) and an adult stage. Juvenile *Meloidogynes* parasites hatch from eggs as vermiform, second-stage juveniles (J2), the first moult having occurred within the egg. Newly hatched juveniles have a short free-living stage in the soil, in the rhizosphere of the host plants. They may reinvade the host plants of their parent or migrate through the soil to find a new host root. J2 larvae do not feed during the free-living stage, but use lipids stored in the gut.

Plate 23: ***M. incognita*** **on Root.**

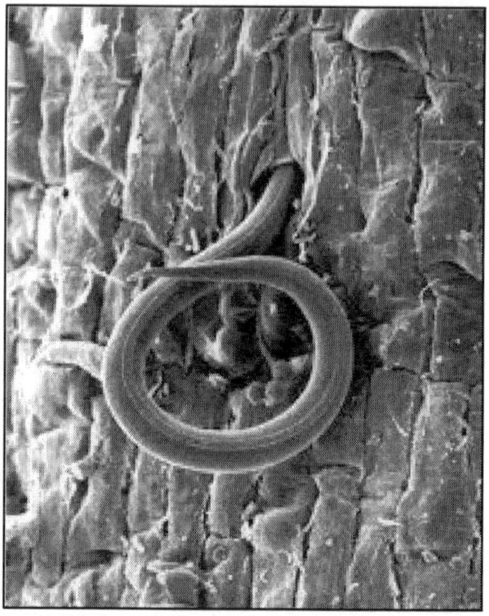

Plate 24: ***M. incognita*** **on Root.**

An excellent model system for the study of the parasitic behaviour of plant-parasitic nematodes has been developed using*Arabidopsis thaliana* as a model host. The *Arabidopsis* roots are initially small and transparent, enabling every detail to be seen. Invasion and migration in the root was studied using *M. incognita*. Briefly, second stage juveniles invade in the root elongation region and migrate in the root until they became sedentary. Signals from the J_2 promote parenchyma cells near the head of the J_2 to become multinucleate to form feeding cells, generally known as giant cells, from which the J_2 and later the adults feed.Concomitant with giant cell formation, the surrounding root tissue gives rise to a gall in which the developing juvenile is embedded. Juveniles first feed from the giant cells about 24 hours after becoming sedentary.

After further feeding, the J_2s undergo morphological changes and become saccate. Without further feeding, they moult three times and eventually become adults. In females, which are close to spherical, feeding resumes and the reproductive system develops. The life span of an adult female may extend to three months, and many hundreds of eggs can be produced. Females can continue egg laying after harvest of aerial parts of the plant and the survival stage between crops is generally within the egg.

The length of the life cycle is temperature-dependent. The relationship between rate of development and temperature is linear over much of the root-knot nematode life cycle, though it is possible the component stages of the life cycle, *e.g.* egg development, host root invasion or growth, have slightly different optima. Species within the *Meloidogyne*genus also have different temperature optima. In *M. javanica,* development occurs between 13 and 34 °C, with optimal development at about 29 °C.

Gelatinous Matrix

Root-knot nematode females lay eggs into a gelatinous matrix produced by six rectal glands and secreted before and during egg laying. The matrix initially forms a canal through the outer layers of root tissue and later surrounds the eggs, providing a barrier to water loss by maintaining a high moisture level around the eggs. As the gelatinous matrix ages, it becomes tanned, turning from a sticky, colourless jelly to an orange-brown substance which appears layered.

Egg Formation and Development

Egg formation in *M. javanica* has been studied in detail, and is similar to egg formation in the well studied, free-living nematode *Caenorhabditis elegans*. Embryogenesis has also been studied, and the stages of development are easily identifiable with a phase contrast microscope following preparation of an egg mass squash. The egg is formed as one cell, with two-cell, four-cell and eight-cell stages recognisable. Further cell division leads to the tadpole stage, with further elongation resulting in the first stage juvenile, which is roughly four times as long as the egg. The J_1 stage of *C. elegans*has 558 cells, and the J_1 of *M. javanica* likely has a similar number, since all nematodes are morphologically and anatomically similar. The egg shell has three layers, with the vitelline layer outermost, then a chitinous layer and a lipid layer innermost.

Egg Hatching

Preceded by induced changes in eggshell permeability, hatching may involve physical and/or enzymatic processes in plant-parasitic nematodes. Cyst nematodes, such as *Globodera rostochiensis*, may require a specific signal from the root exudates of the host to trigger hatching. Root-knot nematodes are generally unaffected by the presence of a host, but hatch freely at the appropriate temperature when water is available. However, in an egg mass or cyst, not all eggs will hatch when the conditions are optimal for their particular species, leaving some eggs to hatch at a later date. Ammonium ions have been shown to inhibit hatching and to reduce the plant-penetration ability of *M. incognita* juveniles that do hatch.

Reproduction

Root-knot nematodes exhibit a range of reproductive modes, including sexuality (amphimixis), facultative sexuality, meiotic parthenogenesis (automixis) and mitotic parthenogenesis (apomixis).

Vegetables

M. incognita is a major pest of vegetables throughout the tropics and subtropics. Many crops grown as vegetables are susceptible to the nematode particularly tomato, aubergine, okra, cucumber, melon, carrot, gourds, lettuce and peppers. Estimates of vegetable crop losses due Meloidogyne species, mainly *M. incognita* and *M. javanica*, have ranged from 17 to 20 per cent for aubergine (*Solanum melongena*), 18 to 33 per cent for melon and 24 to 38 per cent for tomato. In India, yield losses of okra (*Abelmoschus esculentus*), tomato and aubergine field crops are estimated at 91 per cent, 42 to 54 per cent and 18 per cent, respectively. Other trials in India using nematicides showed that in fields naturally infested with mixed populations of *M. incognita* and *M. javanica* the avoidable yield losses in peas, okra, tomato and bottle gourd (*Lagenaria siceraria*) crops is 46 per cent to 56 per cent.

Yams

In India, pot experiments have demonstrated that even low nematode populations of 100 M. *incognita* juveniles per plant can cause a highly significant reduction in yam (*D. rotundata*) tuber weights of 27 per cent and larger nematode populations over 55 per cent yield loss (Mohandas and Ramakrishnan,1997).

Potatoes

Losses of potatoes due to Meloidogyne species, mainly *M. incognita*, are estimated at 25 per cent or more.

Spices

M. incognita is an important pest of black pepper (*Piper nigrum*) in many countries particularly in Brazil and India where infestation levels of over 90 per cent have been reported (Koshy and Bridge, 1990). As few as 10 juveniles per plant can reduce black pepper growth by 16 per cent under potted conditions (Koshy *et al.*, 1979). Cardamom (*Elettaria cardamomum*) suffers yield losses of 32 to 47 per cent in the presence of *M. incognita* in India (Ali, 1986). In Australia, *M. incognita* and *M. javanica* are pests of ginger and severe infestation of rhizomes can reduce yields by 57 per cent (Pegg *et al.*, 1974). In pot experiments, a reduction of 74 per cent in ginger rhizome weight has been recorded with initial *M. incognita* soil populations of 10,000 nematodes.

Coffee

M. incognita is found in many coffee-growing areas in the world but it is in Brazil where its effects have been the most catastrophic. In Brazil, the nematode has caused the complete destruction of coffee plantations and resulted in major agricultural changes as farmers were forced to adopt alternative crops.

Prevention and Control

- ☆ Methods for the control of *M. incognita* vary with the production system used and the value of the crop. Chemical control may be used on high-value crops, and a wide range of chemicals are available. Strategies of cultural control are less well developed and crop rotations are difficult to design because of the wide host range of *M. incognita*. Groundnuts or maize, which are both poor or non-hosts to *M. incognita*, have been evaluated for use in cropping systems designed to manage this nematode. Resistance to M. incognita exists in a number of commercial crop varieties, principally vegetables.

2. Cereal Cyst Forming Nematode

Molya Disease of Wheat and Barley

Heterodera avenae **Wollenweber 1924**

Heterodera major

Kingdom: Animalia

Phylum: Nematoda

Class: Secernentea

Subclass: Diplogasteria

Order: Tylenchida

Superfamily: Tylenchoidea

Family: Heteroderidae

Subfamily: Heteroderinae

Genus: *Heterodera*

Species: *H. avena*

Introduction

Heterodera avenae, the **cereal cyst nematode** or **Ustinov cyst nematode**, is a plant pathogen and an obligate parasite of cereal crops including barley, oats, wheat and rye. Cereal crops infected with this nematode are more susceptible to infection by fungal diseases such as rhizoctonia root rot.

Economic Significance

Infection with cereal cyst nematode is associated with a reduction in grain yield which may be of significant proportions. Thresholds of harmfulness depend on the population density of the nematode and differ among various cultivars and varieties. The weather conditions and soil types are also an important factor, with infections being increased in damp, warm growing seasons. The symptoms of infection include stunting and yellowing of the visible parts of the plant which may give the crop an uneven appearance. The presence of the disease can be established by inspecting the root system of plants and looking for abnormal development. The roots of infested plants develop a tangle of branches and swellings which are white when young but turn dark brown with age. It is the seedlings of cereals that are most readily infected with this nematode and damaged roots are often invaded by soil-borne pathogens such as root and crown rots. When nematode populations are high, significant losses in yield can occur. All wheat varieties are susceptible but some cultivars do not support cyst formation. It was found in trials that direct-drilling that did not disturb the soil below seeding depth reduced the incidence of cereal cyst nematode but did not have the same effect on the other serious cereal seedling diseases of rhizoctonia root rot and take-all. Control is by crop rotation, fallowing and the use of less susceptible varieties.

Host Plants/Species Affected

Avena sativa (oats)

Hordeum vulgare (barley)

Poaceae (grasses)

Secale cereale (rye)

Triticale

Triticum (wheat)

Triticum aestivum (wheat)

Zea mays (maize)

Identification

- ✰ This species is a cyst nematode. Fertilized females are white to brown and of the size of a pinhead
- ✰ Eggs are contained in the cysts.

Damage

In spring, patches of stunted vegetation can be observed. In June, affected plants exhibit very short and shrubby roots. Conditions favourable to the appearance of damage are periods of humidity in winter or in spring, followed by dryness in late spring.

Impact

H. avenae is a major pest of wheat, barley and oats. It is a major biotic constraint to cereal production in north-western India and southern Australia. The annual loss due to the cereal cyst nematode in wheat was estimated at $70 million in Australia, US$4.5 million in Europe, and US$9 million in India. In the USA, *H. avenae* is ranked sixteenth on the basis of its expected economic impact among exotic pests, with expected losses of US$312 million.

In Australia, a population of two eggs per g of soil is considered the threshold level for wheat. An initial population of two eggs per g of soil reduces wheat yields by 20 per cent, and at four eggs per g of soil the loss is 35 per cent. In temperate climates, the threshold levels for oats, wheat and barley are, respectively, 0.2, one and five eggs + juveniles per gramme of soil. The degree of damage depends on initial population, type of soil, climatic conditions, crop species and cultivar, and interaction with secondary pathogens.

List of Symptoms/Signs

Leaves - abnormal colours

Roots - cysts on root surface

Stems - stunting or rosetting

Whole plant - dwarfing

Whole plant - early senescence

Symptoms

Infested fields show uneven patches of stunted plants randomly distributed throughout the field. Heavily infected plants have reduced tillering. Cereal heads produced by stunted plants are usually small and poorly filled.

Wheat

The leaves of heavily infected plants are reddish-yellow with thin and narrow leaf blades. Ears, if formed, have very few grains. The root system shows a characteristic elongation of the main root. The tips of the rootlets appear bunched and slight swellings may be present at the point of cyst attachment. In some cases individual roots have a knotted appearance with several females at each knot **(Plates 25 and 31).**

Barley

In fields, there are patches of stunted plants with fewer tillers. Leaf tips become yellow. Roots are generally not much affected; however, excessive branching has been observed **(Plate 32).**

Oats

Oat leaves become red. Roots are shortened and thickened by the nematode infection **(Plate 26).**

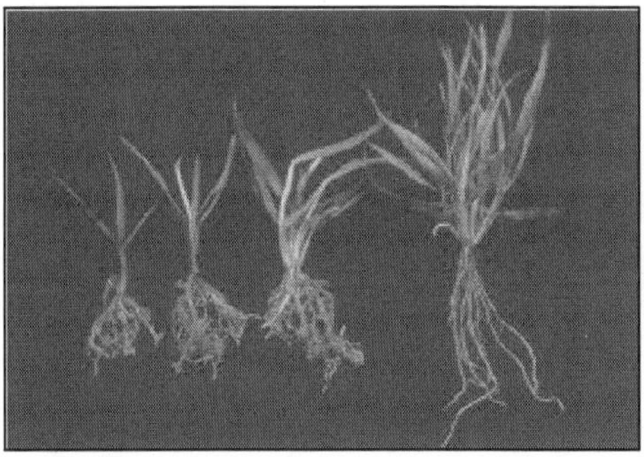

Plate 25: *Heterodera avenae* Damage.

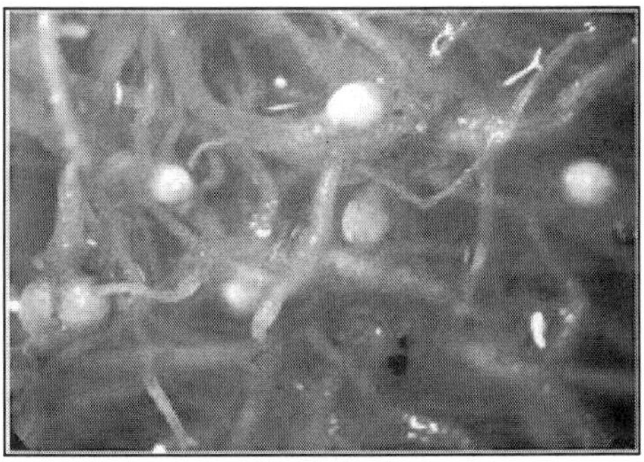

Plate 26: White Female of *Heterodera avenae.*

Plate 27: Oat Cyst Nematode.

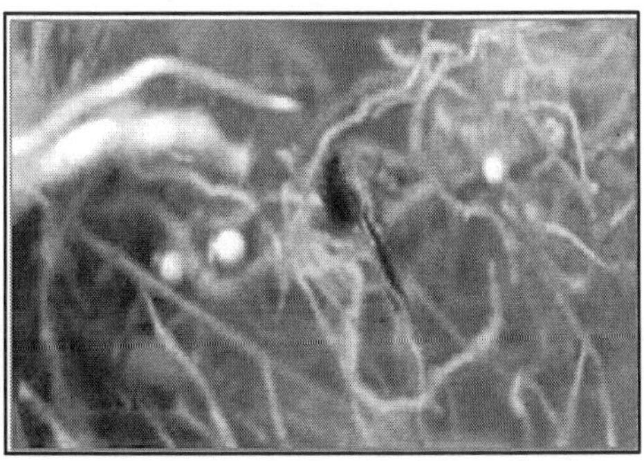

Plate 28: Cereal Cyst Nematode.

Plate 29: Soybean Cyst Nematode.

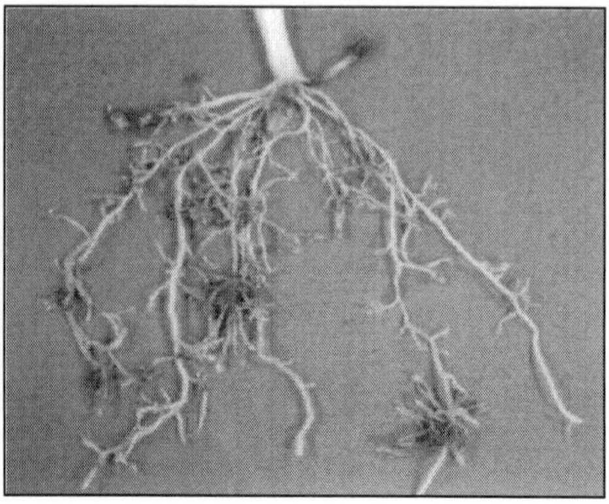

Plate 30: Cereal Cyst Nematode.

Plate 31: Cyst Nematode Damage on Wheat Field .

Plate 32: Cyst Nematode Damage on Spring Barley.

Biology

Females develop on roots and change into cysts, which are eventually released into the soil. These enable the eggs to survive many years until new host plants stimulate eggs and allows the nematodes to hatch, to develop and, eventually, reproduce **(Plate 27-32).**

Life Cycle

This microscopic nematode exhibits sexual dimorphism. The female is rounded and white and measures 680 by 930 micrometres. The male is vermiform and transparent and measures 40 by 1300 micrometres. The eggs are oval and the vermiform larvae moult four times. The second instar larvae are mobile and can travel distances of up to thirty centimetres, looking for and invading roots of suitable host species. Here they develop, growing into sedentary bottle-shaped third instar larvae and rounded fourth instar ones. These then develop into either females or males and mating takes place. The female retains the majority of the several hundred eggs she produces inside her body. She turns into a brown cyst by the end of plant growth season, as her external surface hardens and her internal parts die. The larvae may remain in anabiosis within this cyst for several years until suitable host plants become available. The cysts can withstand harsh conditions and can be spread in soil, by agricultural machinery, by animals, by wind, by dust storms or other means.

Prevention and Control

Cultural Control and Sanitary Methods

- ☆ Crop rotation is highly effective in bringing the population levels down. Three-year rotations with non-host crops such as legumes greatly reduce nematode populations and also benefit nitrogen-depleted soils. Rotation with crucifers is also effective. The rotation fallow/wheat/legume pasture is widely practised.
- ☆ The application of fertilizers enhances the tolerance of cereals to *H. avenae* and urea is used to raise soil nitrogen levels in India and Australia. However, fertilizer application at higher rates may also be responsible for increased numbers of cysts.
- ☆ Agricultural practices such as fallowing and summer ploughing are also highly effective in reducing populations of *H. avenae*. Fallowing checks population build-up by starvation of emerged juveniles; however, this method is unattractive because of reduction of the cropping period available to farmers and the need for thorough weed control. Repeated deep summer ploughing in hot climates reduces the incidence of *H. avenae*, but may cause wind erosion and cyst dispersion in lighter soils.

Host-Plant Resistance

The use of resistant cultivars is the most promising approach for controlling *H. avenae*. Resistant plants allow penetration of second-stage juveniles but resist the development of females, thus inhibiting proliferation. Many efficient sources of resistance to cereal cyst nematode are present in various species and cultivars of Hordeum, Avena, Triticum, Secale and Triticale. These sources of resistance are not always efficient against all pathotypes. It is always important to identify the nematode pathotype before choosing the resistant cereal cultivar.

Biological Control

H. avenae females and eggs are often parasitized by a range of fungi in many soils, of which *Nematophthora gynophila* and *Verticillium chlamydosporium* are the most important. The fungi parasitize females on roots and sometimes destroy them completely or cause formation of smaller cysts. Parasitized females have lower fertility and may be fragile. *N. gynophila* is widespread in the UK and V. chlamydosporium is widely disseminated in southern Australia. In soils where cereals have been grown for a long time, natural suppression of cereal cyst nematode has been observed which is largely attributed to fungal parasites. Natural control requires at least 5 years building up antagonist populations for the control of *H. avenae*. The natural population levels can be augmented by altering rotation systems and by introducing artificial inocula. The bacterial parasites *Pasteuria penetrans* and *Pasteuria* sp. have been reported from India and the UK, respectively.

Chemical Control

- ✰ Application of DD chemical @ 300 L/ha as soil fumigation or Nemagon @45 L/ha.
- ✰ Soil treatment with carbofuran @ 2kg a.i./ha.

3. Ear Cockle (Seed Gall) Disease of Wheat

(*Anguina tritici* (Steinbuch, 1799) Filipjev, 1936)

Classification

Tylenchida

Tylenchina

Tylenchoidea

Anguinidae

Anguininae

Anguina tritici

History and Significance

Anguina tritici was the first plant parasitic nematode to be described in the literature in 1743. It causes a disease in **wheat** and **rye** called "ear-cockle" or seed gall. Originally found in many parts of the world but has been eradicated from the western hemisphere. Currently in north Africa and west Asia. Yield loses up to 70 per cent have been reported, ranging from 30-70 per cent. Threshold of 10,000 juveniles/kg soil develop disease.

Morphology and Anatomy

It is a large nematode, ranging from 3–5 mm in length. *Anguina tritici* has a three part esophagus and the esophageal glands do not overlap with intestine. The female body tends to be thickened and curved ventrally. It has a short stylet (8-11 μm). Females have one ovary and the vulva located posterior. Males possess small spicules and small bursae or alae **(Plate 33).**

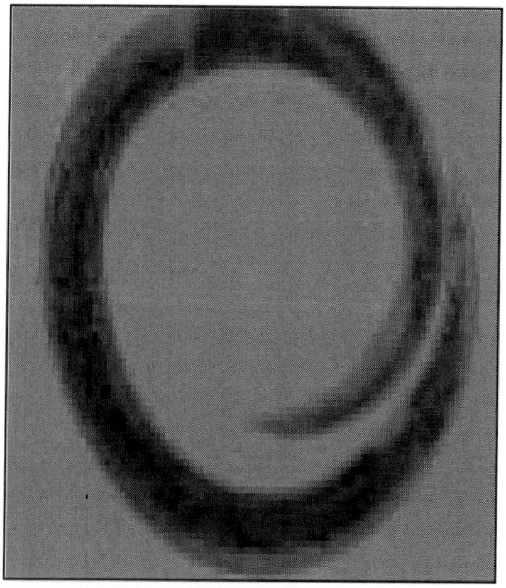

Plate 33: Morphology and Anatomy of *A. tritici*.

Distribution

India, Ethiopia, Syria, Rumania, Yugoslavia. Nematode occurs throughout the world and is easily disseminated in seed. Anguina tritici may be considered a "museum relic" nematode in the western world. Up to 9 per cent of wheat samples in England between 1918-1936 contained galls. From 1949-1957, 0-0.02 per cent contained galls; none found since 1957.

Feeding

Infective stage, J2 does not parasitize roots, but moves on the root and stem surface in a film of water to the stem growing tip. Juveniles probably feed ectoparasitically between compacted leaves until flower primordia form. The ectoparasitic feeding can result in leaf curling in the host.

Hosts: Wheat and rye.

Life Cycle

Juveniles (J2) penetrate flower primordia and develop through the third and fourth stages to adulthood. The final molt to adulthood occurs only after the seed gall has formed. Galls can develop from undifferentiated flower buds, stamen tissues, and various other tissues. Galls contain up to 80 adults in a 1:1 sex ratio. Reproduction is amphimictic; mating occurs and females produce up to 2000 eggs per individual over several weeks. Thousands of J2s are present in seed galls. Galls fall to the ground, absorb water, and release juveniles in springtime, or galls may be harvested and stored with seed. Juveniles within drying galls can enter a cryptobiotic state; viable juveniles have been recovered for up to periods as long as 38 years. Galls appear darker, shorter, and thicker than seed kernels. When galls become wet and absorb water, juveniles are activated. For nematodes forming seed galls, one generation is produced per year; more than one generation per year can be produced for nematodes that form leaf galls and more stages may be found in the gall.

Damage (Plate 34-37)

Ectoparasitic feeding of *A. tritici* may cause leaf rolling, curling, and spiralling. Plants mature more slowly, and produce smaller seed heads. Yield depressions of 50 per cent are common.

Crop losses occur frequently in areas with primitive agricultural practices, because seed flotation, mechanical separation, and crop rotation are not practiced.

The interaction of this nematode and *Clavibacter tritici* in wheat in India results in an oozing bacterial infection of the grain known as **Tundu disease**. The disease is also known as spike blight or yellow ear rot.

The bacterium appears unable to cause the disease without the nematode. The bacterium is vectored on the surface of the nematode into the inflorescence. It colonizes the galled tissue stimulated by the nematode, and out-competes the nematode for this resource (Bird).

Host Plant Relationship

It is an ectoparasite that becomes endoparasitic invading inflorescence and developing seeds. It causes a disease called "ear-cockle", "gout" or seed gall on wheat and rye. It is not a host of oat, maize and shorgum. On wheat it causes stunted plants, distorted leaves; seeds are transformed into galls which contain a dried mass

Plate 34: Wheat Seed-Gall, Nematode.

Plate 35: Ectoparasitic Feeding of *A. tritici* on Wheat.

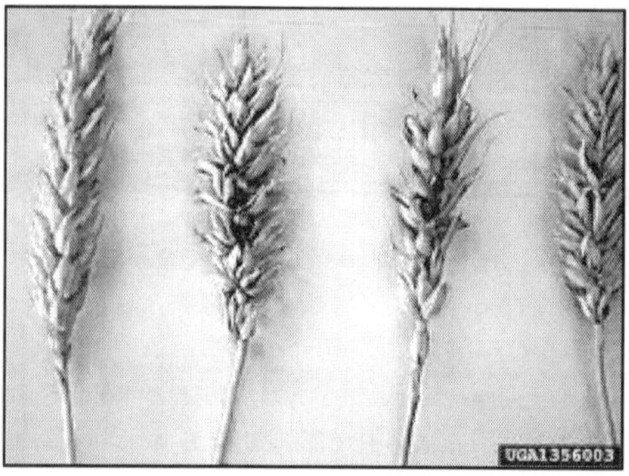

Plate 36: Wheat Seed-Gall Nematode on Common Wheat.

Plate 37: Wheat Gall Nematode.

of nematodes. If compared to normal wheat seeds, galls are smaller in size, lighter, and their color ranges from light brown to black (normal wheat seeds are tan in color).

Management

- ☆ Crop rotation for 1 to 2 years to a non-host eliminates *A. tritici* from the soil. The nematode does not survive by feeding on fungi.
- ☆ Seed can be cleaned by placing it in a 20 per cent brine solution; galls float to the surface where they can be separated. The seed is then rinsed and dried (heat treated).
- ☆ Mechanical separation also effective in removing galls from seed.
- ☆ There are no resistant varieties of wheat. However, there are resistant varieties of barley.

Chapter 3
Centipedes and Millipedes

General Information

Classification

Kingdom: Animalia

Phylum: Mandibulata

Class: Chilopoda (centipedes)

Diplopoda: (millipedes)

Centipedes and millipedes are not insects because they have more than six legs, but they are closely related invertebrates. When outdoors, these invertebrates are innocuous organisms, but they may be considered pests when they share living space with us. They are most often found in moist habitats or areas with high humidity. Both of these groups of invertebrates have long, segmented bodies with either one pair (centipedes) or two pairs (millipedes) of legs on each segment. Their food preferences vary greatly.

Millipedes, and Centipedes in the Home

They are not harmful to food, clothes, furniture, or other items within homes, although just their presence can be disturbing. Their preferred habitat is moist, decaying leaf litter or other organic material found around building foundations. When they are found outdoors, we consider them a beneficial group of animals.

During late summer and fall and occasionally during spring and summer, millipedes can leave the soil and leaf litter and crawl into homes, sometimes in very large numbers. They enter through cracks in foundations, around ground-level windows, and under doors. They are commonly found in basements although they may also be found in ground-level rooms. They enter for one of two reasons. First, in the fall, millipedes and sowbugs seek protected places to overwinter. Second, excessive rainfall or ground moisture may force them out of their normal environment in the soil to areas with less moisture. Millipedes and sowbugs can also be found indoors in early spring as they emerge from cracks and crevices where they spent the winter. Sowbugs and millipedes are rarely seen indoors during winter.

It is common for centipedes to actively move indoors from outdoor harborage areas during spring and summer. They are most commonly seen in homes during warm weather. They can be found indoors during winter but are less common. Millipedes and centipedes are often seen in and around gardens and may be found wandering into homes. Unlike insects, which have three clearly defined body sections and three pairs of legs, they have numerous body segments and numerous legs. Like insects, they belong to the largest group in the animal kingdom, the arthropods, which have jointed bodies and legs and no backbone. Their bodies are covered with a shell-like covering called an exoskeleton. There are several features that distinguish a millipede from a centipede.

Millipedes and centipedes do not carry diseases that affect people, animals or plants. Millipedes do occasionally damage seedling plants by feeding on stems and leaves, and may enter homes in large numbers during periods of migration and become a considerable nuisance. They do not cause damage inside the home, although they may leave a stain if they are crushed. Centipedes, which have poison glands and can bite, pose an occasional threat to humans.

Differences between Centipedes and Millipedes

Characteristic	*Centipede*	*Millipede*
Class	Chilopoda	Diplopoda
Antennae	Long	Short
Body	Flexible, dorsoventrally flattened arthropods	More rigid arthropods distinguishable by their subcylindrical shape
Number of Legs	One pair per body segment	Two pairs per body segment, except for the first three segments, which have one pair each
Appearance of Legs	Visibly extend from sides of body; trail backward behind body	Do not visibly extend from body; rear leg pairs in line with body
Role of legs	Centipedes have maxillipeds to attack prey. These are the venomous modified legs of the centipede's first segment and can be used in defense	In the absence of stinging structures, millipedes employ defensive secretions to protect themselves from predators
Movement	Fast runners	Slow walkers
Ability to Bite	Can bite	Do not bite
Feeding Habits	Mostly predatory/carnivores and kill their prey by injecting them with venom	Mostly scavengers/primarily detritivores/feed primarily on decaying organic matter and they may eat the roots and leaves of seedling plants

Common Characteristics of Cetipedes and milipedes

- ☆ Have long worm-like body with many segments
- ☆ Have many pairs of jointed legs; centipedes have one pair per body segment and millipedes have two pairs per body segment
- ☆ Body covered by a flexible outer skeleton called an exoskeleton
- ☆ Both breathe through spiracles and have no direct copulatory organs.
- ☆ **Centipedes have a pair of poison claws on first body segment, or head**; millipedes do not have poison claws.

Centipedes

Centipedes (from Latin prefix ***centi-***, "**hundred**", and ***pes, pedis, "foot"***) are arthropods belonging to the class **Chilopoda** of the subphylum Myriapoda. They are elongated metameric creatures with one pair of legs per body segment. Despite the name, centipedes can have a varying number of legs, ranging from 30 to 354. Centipedes always have an odd number of pairs of legs. Therefore, no centipede has exactly 100 legs. A key trait uniting this group is a pair of venom claws or forcipules formed from a modified first appendage. Centipedes are predominantly carnivorous. Worldwide, an estimated 8,000 species of centipedes are thought to exist, of which 3,000 have been described. Great variation exists among the 3,000 species of centipedes. Not all of them have 100 legs – some have more and others have fewer. Some are short and others are long. Some are blind, while others boast multiple eyes. Centipedes rely on their antennae to help them get along **(Plates 38-39)**

Their size can range from a few millimetres in the smaller lithobiomorphs and geophilomorphs to about 30 cm (12 in) in the largest scolopendromorphs. Centipedes can be found in a wide variety of environments. They normally have a drab coloration combining shades of brown and red. Cavernicolous (cave-dwelling) and subterranean species may lack pigmentation, and many tropical scolopendromorphs have bright aposematic colours.

The most common centipede is called the house centipede. When full grown, it is more than an inch long. It has a flattened body with 15 pairs of long, jointed legs attached along the sides, one pair of legs per body segment. The legs are very evident

Plate 38: Centipedes.

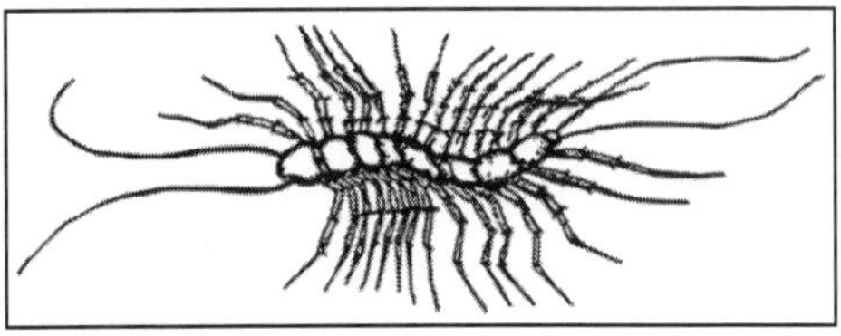

Plate 39: House Centipede.

when the centipede is standing or running. The body is brownish or grayish-yellow sometimes marked with three dark longitudinal stripes visible from above. The legs are clearly striped. A pair of long slender feelers extends forward from the head.

Centipedes move quickly and are usually noticed running across a wall, ceiling, or open room toward a dark area where it is difficult to find them. It is also common for centipedes to stop abruptly and remain motionless before they suddenly begin running again. Centipedes are often associated with damp conditions.

Centipedes are predators, feeding on small, living creatures such as insects, spiders, and other arthropods, like sowbugs, and millipedes. They use poison-filled jaws to help subdue their prey. Because of their predaceous habits, centipedes are beneficial arthropods.

Bites by a centipede are rare because it is very shy. Its jaws are also small and cannot break through human skin easily unless the centipede is pressed or squeezed. In the rare event of a bite, some swelling may be expected, but thc pain should not be severe.

Habitat

Centipedes have a wide geographical range, where they even reach beyond the Arctic Circle. They are found in an array of terrestrial habitats from tropical rainforests to deserts. Within these habitats, centipedes require a moist microhabitat because they lack the waxy cuticle of insects and arachnids, therefore causing them to rapidly lose water. Accordingly, they are found in soil and leaf litter, under stones and dead wood, and inside logs. Centipedes are among the largest terrestrial invertebrate predators, and often contribute significantly to the invertebrate predatory biomass in terrestrial ecosystems. During the day they hide in dark cracks and crevices, coming out at night to search for insects to eat. House centipedes are actually beneficial, as they capture flies, cockroaches, and other small household pests. They never damage plants or household items. When provoked, a few large kinds of centipedes can inflict a painful bite that may cause localized swelling, discoloration, and numbness.

Description

Centipedes have a rounded or flattened head, bearing a pair of antennae at the forward margin. They have a pair of elongated mandibles, and two pairs of maxillae. The first pair of maxillae form the lower lip, and bear short palps. The first pair of limbs stretch forward from the body to cover the remainder of the mouth. These limbs, or maxillipeds, end in sharp claws and include venom glands that help the animal to kill or paralyze its prey.

Many species of centipedes lack eyes, but some possess a variable number of ocelli, which are sometimes clustered together to form true compound eyes. However, these eyes are only capable of discerning light and dark, and have no true vision. In some species, the first pair of legs at the head end of the centipede acts as sense organs similar to antennae, but unlike the antennae of most other animals, theirs point backwards. Unusual sense organs found in some groups are the **organs of Tömösváry**. These are located at the base of the antennae, and consist of a disc-like structure with a central pore surrounded by sensory cells. They are probably used for sensing vibrations, and may even provide a sense of hearing.

Forcipules are a unique feature found only in centipedes and in no other arthropods. The forcipules are modifications of the first pair of legs, forming apincer-like appendage always found just behind the head. Forcipules are not true mouthparts, although they are used in the capture of prey items, injecting venom and holding onto captured prey. Venom glands run through a tube almost to the tip of each forcipule.

Behind the head, the body consists of 15 or more segments. Most of the segments bear a single pair of legs, with the maxillipeds projecting forward from the first body segment, and the final two segments being small and legless. Each pair of legs is slightly longer than the pair immediately in front of it, ensuring that they do not overlap, so reducing the chance that they will collide with each other while moving swiftly. In extreme cases, the last pair of legs may be twice the length of the first pair. The final segment bears a telson and includes the openings of the reproductive organs.[6]

As predators, centipedes mainly use their antennae to seek out their prey. The digestive tract forms a simple tube, with digestive glands attached to the mouthparts. **Like insects, centipedes breathe through a tracheal system,** typically with a single opening, or spiracle on each body segment. They excrete waste through a single pair of malpighian tubules.

Scolopendra gigantea, also known as the **Amazonian giant centipede**, is the largest existing species of centipede in the world, reaching over 30 cm (12 in) in length. It is known to eat lizards, frogs, birds, mice, and even bats, catching them in midflight, as well as rodents and spiders.

Life Cycle

Adult centipedes overwinter in secluded moist places. Eggs are usually placed in damp soil in the spring and summer. Some centipede species add segments and legs as they grow; others are born with a complete set. Centipedes require two to three years to mature and have been known to live six years. Development is slow, with some species developing through as many as 10 instars, or molts. A few species give birth to living young. Centipedes are relatively long-lived; some species have been known to live up to six years.

The house centipede, *Scutigera coleoptrata*, is the only species capable of reproducing in homes and is often seen in and around homes where dampness occurs. The house centipede is active at night, moving about in search of cockroaches and other insects. Although centipedes rarely bite, are seldom dangerous and are beneficial because they destroy other insects, most people have an aversion to their presence inside their homes.

Centipede reproduction does not involve copulation. Males deposit aspermatophore for the female to take up. In one clade, this spermatophore is deposited in a web, and the male undertakes a courtship dance to encourage the female to engulf his sperm. In other cases, the males just leave them for the females to find. In temperate areas, egg laying occurs in spring and summer, but in subtropical and tropical areas, little seasonality to centipede breeding is apparent. A few species of parthenogenetic centipedes are known.

The Lithobiomorpha and Scutigeromorpha lay their eggs singly in holes in the soil, and the female fills the holes with soil and leaves them. The number of eggs laid ranges from about 10 to 50. Time of development of the embryo to hatching is highly variable and may take from one to a few months. Time of development to reproductive period is highly variable within and among species. For example, it can take 3 years for *S. coleoptrata* to achieve adulthood, whereas under the right conditions, lithiobiomorph species may reach a reproductive period in 1 year. In addition, centipedes are relatively long-lived when compared to insects. For example: the European *Lithobius forficatus* can live for 5 or 6 years. The combination of a small number of eggs laid, long gestation period, and long time of development to reproduction has led authors to label lithobiomorph centipedes as K-selected.

Females of the Geophilomorpha and Scolopendromorpha show far more parental care. The eggs, 15 to 60 in number, are laid in a nest in the soil or in rotten wood. The female stays with the eggs, guarding and licking them to protect them from fungi. The female in some species stays with the young after they have hatched, guarding them until they are ready to leave. If disturbed, the female either abandons the eggs or eats them; abandoned eggs tend to fall prey to fungi rapidly. Some species of Scolopendromorpha are matriphagic, meaning the offspring eat their mother.

Different Orders, Different Sight

Five centipede orders exist:

1. Scolopendromorpha

The Scolopendromorpha, also known as tropical centipedes, possess 21 or 23 body segments (apart from a single species, *Scolopendropsis duplicata*, which has 39 or 43 segments) with the same number of paired legs. Their antennae have 17 or more segments. The eyes have a fixed number of four ocelli on each side in the family Scolopendridae and one ocellus per side in the genus *Mimops* (family Mimopidae), but other families are blind. The order comprises the five families Cryptopidae, Scolopendridae, Mimopidae, Scolopocryptopidae, and Plutoniumidae.

2. Geophilomorpha

The Geophilomorpha, commonly known as soil centipedes, bear upwards of 27 leg-bearing segments. They are eyeless and blind, and bear spiracles on all leg-bearing segments in contrast to other groups, which usually bear them only on their 3rd, 5th, 8th, 10th, 12th, and 14th segments a "mid-body break", accompanied by a change in tagmatic shape, occurring roughly at the interchange from odd to even segments. This group, at 1260 species, the most diverse, also contains the largest and leggiest specimens at 27 or more pairs of legs. They also have 14 segmented antennae. The group includes at least seven families: Mecistocephalidae, Geophilidae (including the former Linotaeniidae, Dignathodontidae and Macronicophilidae), Oryidae, Himantariidae, Schendylidae (including the former Ballophilidae), Zelanophilidae, and Gonibregmatidae (including the former Neogeophilidae and Eriphantidae).

3. Scutigeromorpha

The Scutigeromorpha are anamorphic, reaching 15 leg-bearing segments in length. Also known as house centipedes, they are very fast creatures, and able to withstand falling at great speed: they reach up to 15 body lengths per second when dropped, surviving the fall. They are the only centipede group to retain their original compound eyes, within which a crystalline layer analogous to that seen in chelicerates and insects can be observed. They also bear long and multi-segmented antennae. Adaptation to a burrowing lifestyle has led to the degeneration of compound eyes in other orders; this feature is of great use in phylogenetic analysis.

The group is the sole extant representative of the Notostigmomorpha, defined by having a single spiracle opening at the posterior of each dorsal plate. The more derived groups bear a plurality of spiracular openings on their sides, and are termed the Pleurostigmomorpha. Some even have several unpaired spiracles that can be found along the mid-dorsal line and closer to their posterior section of tergites. There are three families: Psellioididae, Scutigeridae and Scutigerinidae. Pselliodidae includes just a few species in the genus *Sphendononema* (=*Pselliodes*), occurring in the Neotropics and tropical Africa. Scutigerinidae, composed of three species in the genus *Scutigerina*, is restricted to southern Africa and Madagascar. Most scutigeromorphs from other parts of the world belong to the Scutigeridae, which includes two subfamilies, the Scutigerinae and Thereuoneminae.

4. Lithobiomorpha

The Lithobiomorpha, also known as stone centipedes, represent the other main group of anamorphic centipedes; they also reach a mature segment count of 15 trunk segments. This group has lost the compound eyes, and sometimes has no eyes altogether. Instead, its eyes have a single ocellus or a group of ocelli. Its spiracles are paired and can be found laterally. Every leg-bearing segment of this organism has a separate tergite, these alternating in length apart from a pair of long tergites on each of segments 7 and 8. It also has relatively short antennae and legs compared to the Scutigeromorpha. Two families are included, the Henicopidae and Lithobiidae.

5. Craterostigmomorpha

The Craterostigmomorpha are the least diverse centipede clade, comprising only two extant species, both in the genus*Craterostigmus*. Their geographic range is restricted to Tasmania and New Zealand. They have a distinct body plan; their anamorphosis comprises a single stage; they grow from 12 to 15 segments in their first moult. Their low diversity and intermediate position between the primitive anamorphic centipedes and the derived Epimorpha has led to them being likened to the platypus. They represent the survivors of a once diverse clade.

Maternal brooding unites the Craterostigomomorpha with the Epimorpha into the clade Phylactometria. This trait is thought to be closely linked with the presence of sternal pores, which secrete sticky or noxious secretions, which mainly serve to repel predators and parasites. The presence of these pores on the Devonian *Devonobius* permits its inclusion in this clade, allowing its divergence to be dated to 375 (or more) million years ago.

Ecology

Centipedes are predominantly generalist predators, which means they have adapted to eat a variety of different available prey. Examination of centipede gut contents suggests that plant material is an unimportant part of their diets, although centipedes have been observed to eat vegetable matter when starved during laboratory experiments.

Centipedes are mostly nocturnal. Studies on their activity rhythms confirm this, although a few observations of centipedes active during the day have been made, and one species, *Strigamia chinophila*, is diurnal. What centipedes actually eat is not well known because of their cryptic lifestyles and thorough mastication of food. Laboratory feeding trials support that they will feed as generalists, taking almost anything that is soft-bodied and in a reasonable size range. Earthworms may provide the bulk of diets for geophilomorphs, since they burrow through the soil and earthworm bodies would be easily pierced by their venom claws. Geophilomorphs probably cannot subdue earthworms larger than themselves, so smaller earthworms may be a substantial proportion of their diet.

Scolopendromorphs, given their size, are able to feed on vertebrates, as well as invertebrates. They have been observed eating reptiles, amphibians, small mammals, bats, and birds. Springtails may provide a large proportion of lithiobiomorph diets. Little is known about scutigeromorph or craterostigmomorph diets. All centipedes are potential intraguild predators. Centipedes and spiders may frequently prey on one another.

Many larger animals prey upon centipedes, such as mongooses, mice, salamanders, beetles and snakes. They form an important item of diet for many species and the staple diet of some such as the African ant *Amblyopone pluto*, which feeds solely on geophilomorph centipedes, and the South African Cape black-headed snake *Aparallactus capensis*.

Centipede defences include their speed and venomous forcipules, as well as the secretion of defensive chemicals. Geophilomorph centipedes can secrete sticky substances that generate toxic hydrogen cyanide and benzoic acid from microscopic glands on their undersides. Similarly, lithobiomorph centipedes secrete a sticky substance from glands in the rear-most two pairs of legs.

Water regulation is an important aspect of centipede ecology, since they lose water rapidly in dry conditions and are found in moist microhabitats. Water loss is a result of centipedes lacking a waxy covering of their exoskeleton and excreting waste nitrogen as ammonia, which requires extra water. Centipedes deal with water loss through a variety of adaptations. Geophilomorphs lose water less rapidly than lithobiomorphs, though they have a greater surface area to volume ratio. This may be because geophilomorphs have a more heavily sclerotized pleural membrane. Spiracle shape, size, and ability to constrict also have an influence on rate of water loss. In addition, the number and size of coxal pores may be variables affecting centipede water balance.

Centipedes live in many different habitat types–forest, savannah, prairie, and desert, to name a few. Some geophilomorphs are adapted to littoral habitats, where they feed on barnacles. Species of all orders excluding the Craterostigmomorpha have adapted to caves. Centipede densities have been recorded as high as $600/m^2$ and biomass as high as 500 mg/m^2 wet weight. Small geophilomorphs attain highest densities, followed by small lithobiomorphs. Large lithobiomorphs attain densities of $20/m^2$. One study of scolopendromorphs records *Scolopendra morsitans* in a Nigerian savannah at a density of $0.16/m^2$ and a biomass of 140 mg/m^2 wet weight.

Management

Centipedes seldom need to be controlled unless they become a nuisance in the home. Centipedes are predators and generally play a beneficial role in the garden. Their activities should be encouraged in the yard. Reductions in the number of household centipedes occur when their food source–other household pests–is reduced. Airing out damp places may help. Outdoors, centipede control is aided by removing debris as recommended for millipedes.

Millipedes

Evolution

Millipedes are among the first animals to have colonised land during the Silurian geologic period. Early forms probably ate mosses and primitive vascular plants. There are two major groups of entirely extinct millipedes: the Archipolypoda ("ancient, many-legged ones") which contain the oldest known terrestrial animals, and Arthropleuridea, which contain the largest known land invertebrates. The oldest known land creature, *Pneumodesmus newmani*, was a 1 cm (0.39 in) long archipolypodan that lived 428 million years ago in the upper Silurian, and has clear evidence of spiracles (breathing holes) attesting to its air-breathing habits. During the Upper Carboniferous (340 to 280 million years ago), *Arthropleura* became the largest known land invertebrate of all time, reaching lengths of up to 2.6 m (8 ft 6 in). Millipedes also include the earliest evidence of chemical defense, as some Devonian fossils have defensive gland openings called ozopores. Millipedes, centipedes, and other terrestrial arthropods attained very large sizes in comparison to modern species in the oxygen-rich environments of the Devonian and Carboniferous periods, and some could grow larger than one metre. As oxygen levels lowered through time, arthropods became smaller in size.

Introduction

Millipedes are arthropods in the class **Diplopoda**, which is characterised by having two pairs of jointed legs on most body segments. Each double-legged segment is a result of two single segments fused together as one. Most millipedes have very elongated cylindrical or flattened bodies with more than 20 segments, while pill millipedes are shorter and can roll into a ball. Although the name "millipede" derives from the Latin for "thousand feet", no known species has 1,000; the record of 750 legs belongs to ***Illacme plenipes***. There are approximately 12,000 named species classified into sixteen orders and around 140 families, making Diplopoda the largest class of myriapods, an arthropod group which also includes centipedes and other multi-legged creatures.

Most millipedes are slow-moving detritivores, eating decaying leaves and other dead plant matter. Some eat fungi or suck plant fluids, and a small minority are predatory. Millipedes are generally harmless to humans, although some can become household or garden pests, especially in greenhouses where they can cause severe damage to emergent seedlings. Most millipedes defend themselves with a variety of defensive chemicals secreted from pores along the body, although the tiny bristle millipedes are covered with tufts of detachable bristles. Reproduction in most species is carried out by modified male legs called gonopods, which transfer packets of sperm to females.

Millipedes are some of the oldest known land animals, first appearing in the Silurian period. Some members of prehistoric groups grew to over 2 m (6 ft 7 in), while the largest modern species reach maximum lengths of 27 to 38 cm (11 to 15 in). The longest extant species is the giant African millipede (***Archispirostreptus gigas***).

Among myriapods, millipedes have traditionally been considered most closely related to the tiny pauropods, although some molecular studies challenge this relationship. Millipedes can be distinguished from the somewhat similar but only distantly related centipedes (class Chilopoda), which move rapidly, are carnivorous, and have only a single pair of legs on each body segment. The scientific study of millipedes is known as diplopodology, and a scientist who studies them is called a diplopodologist.

Etymology and Names

The scientific name "Diplopoda" comes from the Greek words (*diplous*), "double" and (*podos*), "foot", referring to the appearance of two legs on most segments, as described below. The common name "millipede" is a compound word formed from the Latin roots *mille* ("thousand") and *ped* ("foot"). The term "millipede" is widespread in popular and scientific literature, but among North American scientists, the term "milliped" (without the terminal e) is also used. Other vernacular names include "thousand-legger" or simply "diplopod".

- ☆ Millipedes are similar to centipedes, but have two pairs of legs per body segment.
- ☆ Millipedes are usually brown to blackish in color.
- ☆ The elongated body is rounded, no flattened, and they have no poison claws or legs.
- ☆ They usually coil up when disturbed, similar to the behavior exhibited by sowbugs or pillbugs (a related invertebrate).
- ☆ Millipedes are usually restricted to moist places where they feed on organic matter.
- ☆ In the fall, they may become a nuisance because they migrate away from feeding areas and invade homes.
- ☆ Because they crawl along the ground, they are usually found in lower floors and basements.

- ☆ Once inside the home, they usually die due to desiccation, although in moist basements, they can survive longer.
- ☆ Millipedes live in organic matter (leaves, mulch, piles of wood or wood chips) and other material close to the house.
- ☆ Overmulching and/or overwatering in the garden can result in millipede attack on vegetable plants.
- ☆ Removing the organic debris or mulch materials near your home will help reduce the potential for invading millipedes.

Identification

Millipedes (sometimes called thousand-leggers) are brownish, elongated, and cylindrical to slightly flattened creatures with two (more common) or four pairs of tiny legs per body segment. Millipedes don't really have a thousand legs; even the largest ones have somewhat fewer than a hundred. When they walk, their legs move in an undulating, wavelike manner. Adult millipedes vary from 1/2 to 6 and 1/2 inches in length. When prodded or at rest, most millipedes curl up. The head is rounded with short antennae. They are light brown to black in color (Plates 40-43).

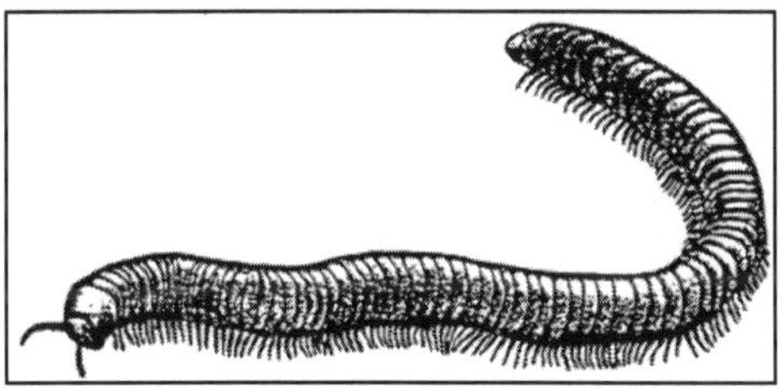

Plate 40: Millipede.

Plate 41: Millipede.

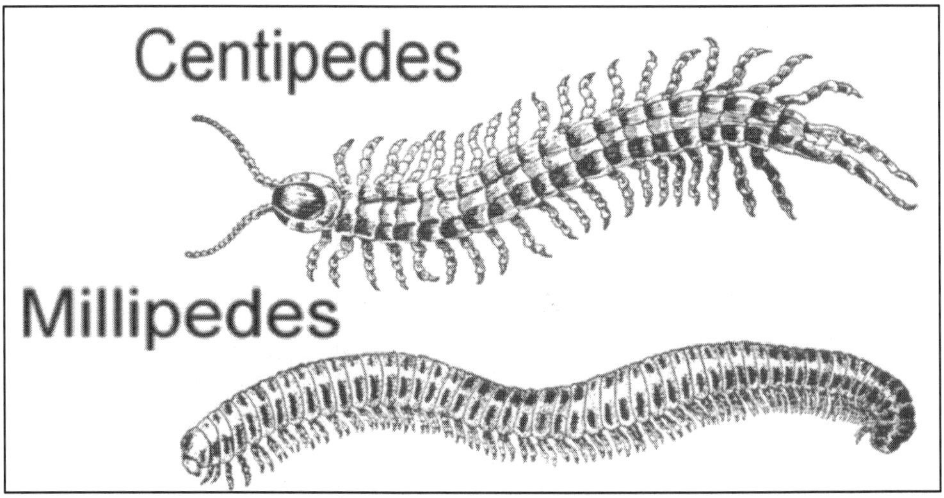

Plate 42: Centipede and Millipede.

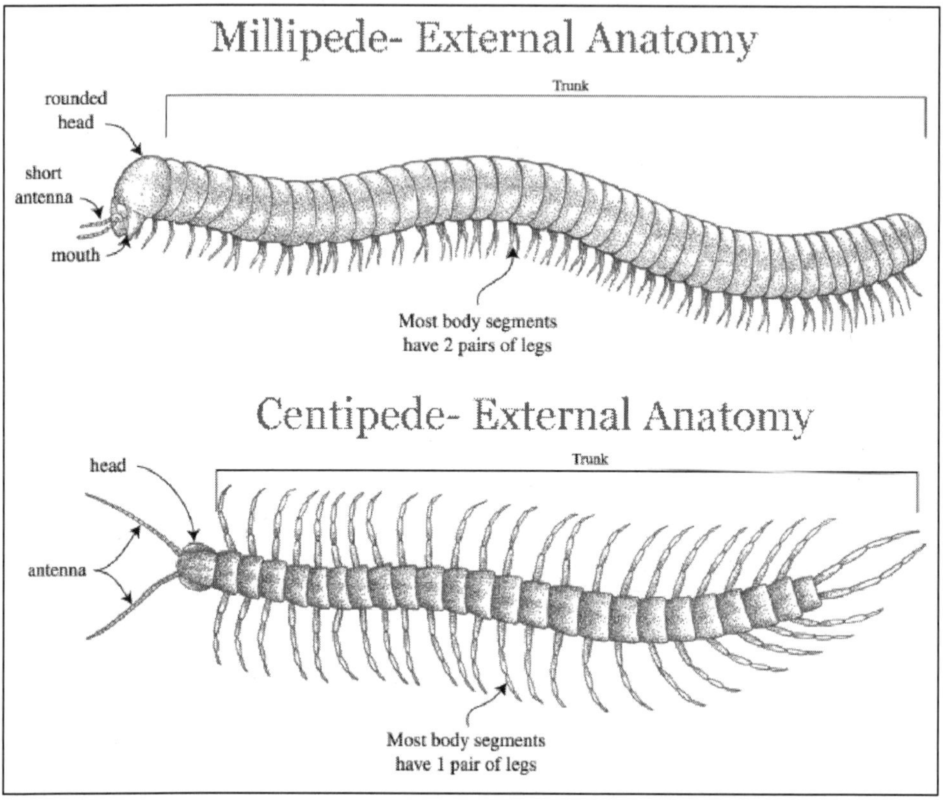

Plate 43: Millipede and Centipede: External Anatomy.

Description

Millipedes are not poisonous, but many species have glands capable of producing irritating fluids that may cause allergic reactions in some individuals. The defensive sprays of some millipedes contain hydrochloric acid that can chemically burn the skin and cause long-term skin discoloration. The fluid can also be dangerous to the eyes. It is not advisable to handle millipedes with your bare hands. Persons handling millipedes may notice a lingering odor on their hands. After contact with millipedes, wash hands thoroughly with soap and water until the odor is gone. The solvents ether or alcohol will also help remove the noxious fluid.

Habitat and Importance

Millipedes normally live in and feed on rotting leaves, rotting wood, and other kinds of moist, decaying plant matter. Generally, their role is a beneficial one in helping to break down dead plant matter. However, when they become numerous, they may damage sprouting seeds, seedlings, or strawberries and other ripening fruits in contact with the ground.

Sometimes individual millipedes wander from their moist living places into homes, but they usually die quickly because of the dry conditions and lack of food.

Occasionally, large numbers of millipedes migrate, often uphill, as their food supply dwindles or their living places become either too wet or too dry. They may fall into swimming pools and drown. When disturbed they don't bite, but some species exude a defensive liquid that can irritate skin or burn the eyes.

Life Cycle

Adult millipedes overwinter in the soil, and eggs are laid in clutches beneath the soil surface. In the spring, millipedes lay between 20 and 300 eggs in the soil. The eggs hatch in several weeks. Young millipedes have only the first three pairs of legs and no more than seven segments. The young grow gradually in size, adding segments and legs as they mature. With each molt, they add additional segments and legs until they reach sexual maturity They mature in two to five years and continue to live for several years thereafter.

Management

- ☆ Millipedes seldom need to be controlled. Keep in mind that they do no damage indoors and pose a minimal health hazard. Those that stray indoors can be swept out or picked up with a vacuum cleaner. Sealing cracks and other openings to the outside helps prevent millipedes from entering. Usually invasions are over within a few days. Eliminating moist hiding places around the home will kill or discourage millipedes. Outdoors, this includes removing rotting wood and decaying grass and leaves from around the house's foundation. This also eliminates millipede food sources. If there is excessive moisture in sub-floor crawl spaces or basements, take measures to dry out these areas. To discourage millipedes in garden areas, reduce mulch and other organic matter and avoid excessive moisture.
- ☆ The first and most important step in controlling millipedes and centipedes is removing or eliminating objects that provide hiding places for the pest outside the home. Eliminate trash piles, rocks, boards, leaf piles, compost piles and similar materials from the immediate vicinity around the outside of the home. Secondly, physically exclude millipedes and centipedes from entering the home where possible. Make sure doors and windows fit tightly and caulk cracks and crevices that permit their entry to the inside. If millipedes or centipedes occur in great numbers or are creating problems, applying pesticide around the building foundation and around doors and windows will provide temporary control.
- ☆ Pesticide may also be applied to mulched flower beds and heavily thatched turf areas where millipedes may be living. Include these areas in a treatment program during periods when millipedes are found inside the home in large numbers. Repeated applications may be necessary to maintain control during periods of heavy millipede migration.

Nonchemical Management

Outdoors

It is very important to make the outside of your home as unfavorable as possible to sowbugs, millipedes, and centipedes to reduce their numbers.

- ☆ Caulk or seal cracks and other openings in exterior foundation walls and around doors and ground-level windows by late summer.
- ☆ Remove leaf litter and decaying vegetation from around the foundation which provide food and shelter for sowbugs and millipedes. A border of bare soil around the building next to the foundation also helps to make the area a less favorable habitat.
- ☆ Trim and thin foundation planting so that ventilation permits the soil to dry more quickly near the foundation.
- ☆ Allow the soil near the house to dry between waterings. Roughening the soil surface will speed drying and will work plant materials into the soil where it is unavailable to sowbugs or millipedes.
- ☆ Reduce thatch in your lawn to discourage sowbugs and millipedes. This is best done by dethatching in early fall.

Indoors

Within homes, sowbugs and millipedes often die quickly because the environment is usually too dry. When sowbugs or millipedes are frequently found alive, it indicates excessive moisture is present. Use a method to dry the room as a long-term solution. Reduce favorable places for centipedes to hide.

- ☆ Remove sowbugs and millipedes with a broom or vacuum cleaner. Kill and remove centipedes as they are seen. Also set out sticky traps (*e.g.*, Roach Motel) on floors where centipedes are often seen to capture them.
- ☆ Remove unnecessary boxes, bags, and other clutter that gives centipedes favorable places to hide.
- ☆ Caulk or seal behind baseboards and in cracks and crevices where centipedes like to hide.
- ☆ Place a dehumidifier in damp areas to sufficiently dry the air. Structural repairs may be necessary to the home and/or yard if a dehumidifier does not keep a room sufficiently dry.

The continued presence of house centipedes suggests that insects are abundant enough to serve as a food supply. Management of insects, spiders, or other prey may be the best treatment for centipedes. Because centipedes feed on a variety or prey, there may not be a single obvious source of food. Reduction of moisture can also help reduce centipede numbers.

Chemical Management

Outdoors

If millipede numbers are still higher than you can tolerate despite these nonchemical steps, supplement your efforts with an insecticide application. Apply a liquid insecticide around the building's foundation and the adjacent ground in a band to help keep pests out. Common insecticides available to the general public for treating building foundations include: bifenthrin, cyfluthrin, deltamethrin, and permathrin. Choose to apply a granular insecticide to the perimeter, such as deltamethrin, lambda-cyhalothrin, or permethrin. Chemical treatment will be less effective if food and shelter exist near the foundation and there are available cracks and spaces for pests to enter the building. Apply an outdoor insecticide in late summer or early fall when sowbugs and millipedes are first noticed indoors.

Indoors

Sowbugs and millipedes often die soon after entering homes because it is too dry, making an insecticide application unnecessary. To treat for centipedes, apply an appropriate insecticide where they hide indoors, especially behind baseboards and in cracks and spaces. Select an insecticide that is labeled for indoor use. Common examples of available insecticides include: bifenthrin, cyfluthrin, deltamethrin, esfenvalerate, and permethrin. These insecticides normally are purchased in ready-to-use aerosol or liquid forms. Insecticides will provide only temporary management if excessive moisture and a food supply exists.

Characteristics

Millipedes come in a variety of body shapes and sizes, ranging from 2 mm (0.079 in) to around 35 cm (14 in) in length, and can have as few as eleven to over one hundred segments. They are generally black or brown in colour, although there are a few brightly coloured species. Body styles vary greatly between major millipede groups. In the basal subclass **Penicillata**, consisting of the tiny bristle millipedes, the exoskelton is soft and uncalcified, and is covered in prominent setae or bristles. All other millipedes, belonging to the subclass **Chilognatha**, have a hardened exoskeleton. The chilognaths are in turn divided into two infraclasses: the **Pentazonia**, containing relatively short-bodied groups such as pill millipedes, and the **Helminthomorpha** ("worm-like" millipedes), which contains the vast majority of species, and the long, many-segmented body types familiar to most people.

Head

The head of a millipede is typically rounded above and flattened below and bears a pair of large mandibles in front of a plate-like structure called a gnathochilarium ("jaw lip").

The head contains a single pair of antennae with seven or eight segments and a group of sensory cones at the tip. Many orders also possess a pair of sensory organs known as the Tömösváry organs, shaped as small oval rings posterior and lateral to the base of the antennae. Their true function is unknown, but they also occur in some centipedes, and are possibly used to measure humidity or light levels in the surrounding environment.

Millipede eyes consist of a number of simple flat-lensed ocelli arranged in a group or patch on each side of the sides of the head. These patches are also called ocular fields or ocellaria. Many species of millipedes, including the entire order Polydesmida and cave-dwelling millipedes such as ***Causeyella*** and ***Trichopetalum***, have secondarily lost their eyes and are completely blind.

Body

Millipede bodies may be flattened or cylindrical, and are composed of numerous metemeric segments, each with an exoskeleton consisting of five chitinous plates: a single plate above (the tergite), one at each side (pleurites), and a plate on the underside (sternite) where the legs attach. In many millipedes, these plates are fused to varying degrees, sometimes forming a single cylindrical ring. The plates are typically hard, being impregnated with calcium salts. Because they lack a waxy cuticle, millipedes are susceptible to water loss and must spend most of their time in moist or humid environments.

The first segment behind the head is legless and known as a collum (from the Latin for neck or collar). The second, third, and fourth body segments bear a single pair of legs each and are known as **"haplosegments"**, from the Greek *haplo*, "single" (the three haplosegments are sometimes referred to as a "thorax"). The remaining segments, from the fifth to the posterior, are properly known as **diplosegments** or double segments. Each diplosegment bears two pairs of legs, rather than just one as in centipedes. This is because each diplosgment is formed by the fusion of two embryonic segments. In some millipedes the last few segments may be legless. The terms "segment" or "body ring" are often used interchangeably to refer to both haplo- and diplosegments. The final segment is known as the telson, which consists of a legless preanal ring, a pair of anal valves (closeable plates around the anus), and a small scale below the anus.

Millipedes in several orders have keel-like extensions of the body-wall known as **paranota**, which can vary widely in shape, size, and texture. Paranota may allow millipedes to wedge more securely into crevices, protect the legs, or make the millipede more difficult for predators to swallow.

The legs are composed of seven segments, and attach on the underside of the body. The legs of an individual are generally rather similar to each other, although often longer in males than females, and males of some species have a reduced or enlarged first pair of legs. The most conspicuous leg modifications are involved in reproduction, and are discussed below.

Internal Organs

Millipedes breathe through two pairs of spiracles located ventrally on each segment near the base of the legs. Each opens into an internal pouch, and connects to a system of tracheae. The heart runs the entire length of the body, with an aorta stretching into the head. The excretory organs are two pairs of malpighian tubules, located near the mid-part of the gut. The digestive tract is a simple tube with two pairs of salivary glands to help digest the food.

Reproduction and Growth

Millipedes show a diversity of mating styles and structures. In the basal order Polyxenida (bristle millipedes), mating is indirect: males deposit spermatophores onto webs they secrete with special glands, and the spermatophores are subsequently picked up by females. In all other millipede groups, males possess one or two pairs of modified legs called gonopods which are used to transfer sperm to the female during copulation. The location of the gonopods differs between groups: in males of the Pentazonia they are located at the rear of the body and known as telopods and may also function in grasping females, while in the Helminthomorpha – the vast majority of species – they are located on the seventh body segment. A few species are parthenogenetic, having few, if any, males.

Gonopods occur in a diversity of shapes and sizes, and in the range from closely resembling walking legs to complex structures quite unlike legs at all. In some groups, the gonopods are kept retracted within the body, while in others they project forward parallel to the body. Gonopod morphology is the predominant means of determining species among millipedes: the structures may differ greatly between closely related species but very little within a species. The gonopods develop gradually from walking legs through successive moults until reproductive maturity.

The genital openings (**gonopores**) of both sexes are located on the underside of the third body segment (near the second pair of legs) and may be accompanied in the male by one or two penes which deposit the sperm packets onto the gonopods. In the female, the genital pores open into paired small sacs called cyphopods or vulvae, which are covered by a small hood-like cover, and are used to store the sperm after copulation. The cyphopod morphology can also be used to identify species. Millipede sperm is aflagellate (lacks a flagellum), a unique trait among myriapods.

In all except the bristle millipedes, copulation occurs with the two individuals facing one another. Copulation may be preceded by male behaviors such as tapping with antennae, running along the back of the female, offering glandular secretions, which the female consumes, or in the case of some pill-millipedes, stridulation or "chirping". During copulation in most millipedes, the male positions his seventh segment in front of the female's third segment, and may insert his gonopods to extrude the vulvae before bending his body to deposit sperm onto his gonopods and reinserting the "charged" gonopods into the female **(Plate 44).**

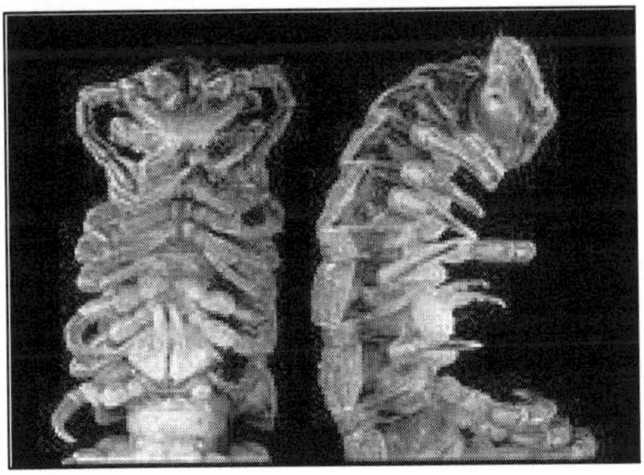

Plate 44: The Gonopods of *Nipponesmus shirinensis.*

Females lay from ten to three hundred eggs at a time, depending on species, fertilising them with the stored sperm as they do so. Many species simply deposit the eggs on moist soil or organic detritus, but some construct nests lined with dried faeces, and may protect the eggs within silk cocoons. In most species, the female abandons the eggs after they are laid, but some species in the orders Platydesmida and Stemmiulida provide parental care for eggs and young.

The young hatch after a few weeks, and typically have only three pairs of legs, followed by up to four legless segments. As they grow, they continually moult, adding further segments and legs as they do so. Some species moult within specially prepared chambers of soil or silk, which they may also use to wait out dry weather, and most species eat the shed exoskeleton after moulting. The adult stage- when individuals become reproductively mature- is generally reached in the final moult stage, which varies between species and orders, although some species continue to moult after adulthood. Furthermore, some species alternate between reproductive and non-reproductive stages after maturity, a phenomenon known as **periodomorphosis**, in which the reproductive structures regress during non-reproductive stages. **Millipedes may live from one to ten years, depending on species.**

Burrowing

The **diplosegments** of millipedes have evolved in conjunction with their burrowing habits, and nearly all millipedes adopt a mainly subterranean lifestyle. They use three main methods of burrowing; bulldozing, wedging and boring. Members of the orders **Julida, Spirobolida** and **Spirostreptida** simply lower their heads and barge their way into the substrate, the collum being the portion of their exoskeleton that leads the way. Flat-backed millipedes in the order **Polydesmida** tend to insert their front end, like a wedge, into a horizontal crevice, and then widen the crack by pushing upwards with their legs, the paranota in this instance constituting the main lifting surface. Boring is used by members of the order **Polyzoniida**. These have smaller segments at the front and increasingly large ones further back; they propel themselves forward into a crack with their legs, the wedge-shaped body widening the gap as they go. Some millipedes have adopted an above-ground lifestyle and lost the burrowing habit. This may be because they are too small to have enough leverage to burrow, or because they are too large to make the effort worthwhile, or in some cases because they move relatively fast (for a milipede) and are active predators.

Diet

The majority of millipedes, are **detritivores** and feed on decomposing vegetation, faeces, or organic matter mixed with soil. They often play important roles in the breakdown and decomposition of **leaf litter:** estimates of consumption rates for individual species range from 1 to 11 percent of all leaf litter, depending on species and region, and collectively millipedes may consume nearly all the leaf litter in a region. The leaf litter is fragmented in the millipede gut and excreted as pellets of leaf fragments, algae, fungi, and bacteria, which facilitates decomposition by the microorganisms. Where earthworm populations are low in tropical forests, millipedes play an important role in facilitating microbial decomposition of the leaf litter. Some millipedes are herbivorous, feeding on living plants, and some species can become serious pests of crops. Millipedes in the order **Polyxenida** graze algae from bark, and **Platydesmida** feed on fungi. A few species are **omnivorous** or occasionally carnivorous, feeding on insects, centipedes, earthworms, or snails. Some species have piercing mouth parts that allow them to feed on plant juices.

Predators and Parasites

Millipedes are preyed on by a wide range of animals, including various **reptiles, amphibians, birds, mammals,** and **insects.** Mammalian predators such as **coatis** and **meerkats** roll captured millipedes on the ground to deplete defensive secretions and rub them off the body before consuming, while certain **poison dart frogs** are believed to incorporate the toxic compounds of millipedes into their own defenses. Several invertebrates have specialized behaviors or structures to feed on millipedes, including larval **glowworm beetles, *Probolomyrmex*** ants, chlamydephorid slugs, and predaceous dung beetles of the genera *Sceliages* and *Deltochilum*. A large subfamily of assassin bugs, the Ectrichodiinae with over 600 species, has specialized in preying upon millipedes.

Parasites of millipedes include **nematodes, phaeomyiid flies**, and **acanthocephalans**.

Defence Mechanisms

Due to their lack of speed and their inability to bite or sting, millipedes' primary defence mechanism is to curl into a tight coil – protecting their delicate legs inside an armoured exoskeleton.

Many species also emit various foul-smelling liquid secretions through microscopic holes called ozopores (the openings of "odoriferous" or "repugnatorial glands"), along the sides of their bodies as a secondary defence. Among the many irritant and toxic chemicals found in these secretions are **alkaloids, benzoquinones, phenols, terpenoids**, and **hydrogen cyanide**. Some of these substances are **caustic** and can burn the exoskeleton of **ants** and other insect predators, and the skin and eyes of larger predators. Primates such as capuchin monkeys and lemurs have been observed intentionally irritating millipedes in order to rub the chemicals on themselves to repel **mosquitoes**. Some of these defensive compounds also show antifungal activity.

The bristly millipedes (order Polyxenida) lack both an armoured exoskeleton and odiferous glands, and instead are covered in numerous bristles that in at least one species, *Polyxenus fasciculatus*, detach and entangle ants.

There are four orders in which six species are of agricultural importane:

1. Spirostreptida
2. Spirobolida
3. Julida
4. Polydesmida

1. Order Spirostreptida

Taxonomy

Kingdom: Animalia

Phylum: Arthropoda

Class: Diplopoda

Subclass: Chilognatha

Infraclass: Helminthomorpha

Superorder: Juliformia

Order: Spirostreptida

Description

Spirostreptida are generally large, long and cylindrical, with 30 to 90 body rings. Eyes are present in most. This order contains the longest millipedes known: the giant African millipedes of the genus *Archispirostreptus* that may exceed 30 centimetres (12 in) **(Plate 45).**

Plate 45: The Giant African Millipedes.

Distribution

Spirostrepitda contains mainly tropical species, and occurs in Africa, Southern Asia to Japan, Australia, and the Western Hemisphere from the United States to Argentina.

2. Order Spirobolida

Taxonomy

Kingdom: Animalia

Phylum: Arthropoda

Subphylum: Myriapoda

Class: Diplopoda

Subclass: Chilognatha

Infraclass: Helminthomorpha

Superorder: Juliformia

Order: Spirobolida

Plate 46: Spriobolidan Millipede.

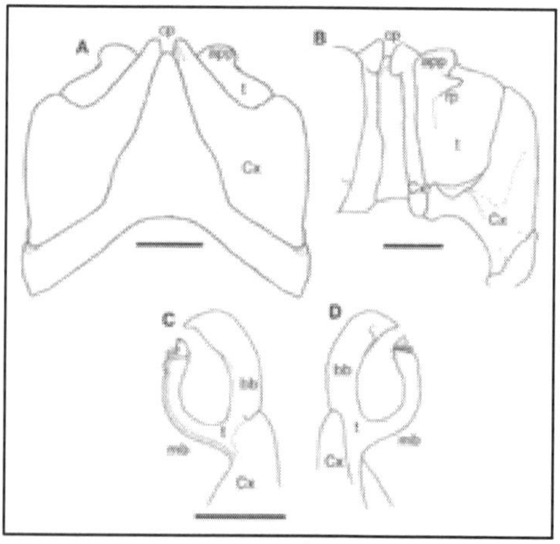

Plate 47: Front and Rear Views of the Anterior (A, B) and Posterior Left (C, D) Gonopods of a Spriobolidan.

Spirobolida is an order of "round-backed" millipedes containing approximately 500 species in 12 families Its members are distinguished by the presence of a "pronounced suture that runs "vertically down the front of the head". Most of the species live in the tropics, and many are brightly coloured. Mature males have two pairs of modified legs, the gonopods, consisting of the 8th and 9th leg pair: the posterior gonopods are used in sperm-transfer while the anterior gonopods are fused into a single plate-like structure **(Plate 46-47)**

3. Order Julida

Taxonomy

Kingdom: Animalia

Phylum: Arthropoda

Subphylum: Myriapoda

Class: Diplopoda

Subclass: Chilognatha

Infraclass: Helminthomorpha

Superorder: Juliformia

Order: Julida

Julida is an order of millipedes. Members are long and cylindrical, typically ranging from 10–120 millimetres (0.39–4.72 in) in length. Eyes may be present or absent, and in mature males of many species, the first pair of legs is modified into hook-like structures. Additionally, both pairs of legs on the 7th body segment of males are modified into gonopods (**Plate 48).**

Plate 48: Julida Millipedes.

Distribution

Julida contains predominantly temperate species ranging from North America to Panama, Europe, Asia north of the Himalayas, and Southeast Asia.

4. Order Polydesmida

Taxonomy

Kingdom: Animalia

Phylum: Arthropoda

Subphylum: Myriapoda

Class: Diplopoda

Superorder: Merocheta Cook, 1895

Order: Polydesmida

Members of the order Polydesmida are also known as "**flat-backed millipedes**", because on most species, each body segment has wide lateral keels known as paranota. These keels are produced by the posterior half (metazonite) of each body ring behind the collum. Polydesmids have no eyes, and vary in length from 3 to 130 mm (0.12 to 5.12 in). Including the telson, adults have 19 or 20 rings, while juveniles may have from 7 to 19 rings. Mature males have a single pair of gonopods consisting of the modified forward leg pair of the 7th segment. Many of the larger species show bright coloration patterns which warn predators of their toxic secretions.

Plate 49: Polydesmid Flat-Backed Millipedes.

Ecology

Polydesmids are very common in leaf litter, where they burrow by levering with the anterior end of the body (**Plate 49).**

Chapter 4
Snails and Slugs

General Information

Snail is a common name that is applied most often to land snails, terrestrial pulmonate gastropod molluscs. However, the common name "snail" is also applied to most of the members of the molluscan class **Gastropoda** that have a coiled **shell** that is large enough for the animal to retract completely into. When the word "snail" is used in this most general sense, it includes not just land snails but also thousands of species of **sea snails** and **freshwater snails.** Occasionally a few other molluscs that are not actually gastropods, such as the **Monoplacophora**, which superficially resemble small limpets, may also informally be referred to as "snails".

Snail-like animals that naturally lack a shell, or have only an internal shell, are mostly called **slugs,** and land snails that have only a very small shell (that they cannot retract into) are often called **semi-slugs**.

The word **"mollusc"** or "mollusk" (both are correct) is derived from the Latin word mollis meaning "soft". The study of molluscs, "malacology", comes from the Greek word for soft, malacos. The term "conchology" is also used for the study of molluscs; however, it is usually applied to those that study the shell only.

Molluscs, in general, are soft-bodied animals that usually produce an external skeleton (called an "exoskeleton") we call a Shell, which is composed of a limey material: calcium carbonate (CaCO3) The shell serves both protective, and supportive purposes. The one feature common to all molluscs is the presence of a fleshy mantle. This is a fold or lobe (or a pair of them) of fleshy material, which secretes, modifies and lines the shell.

Members of all classes except the bivalves possess a ribbon-like set of hooked teeth called radula. These they rasp (think of a fingernail file) back and forth over their food much the same idea as a cat lapping up milk: Vegetarian species use them to scrape algae off rocks and other substrates, while most molluscan carnivores use them to penetrate the surface of their prey - even when that is a decent thickness of shell!

Seven classes of molluscs are recognized:

(1) Gastropoda (Gastropods)

The name means "stomach-footed". (Single shelled cowries, cones etc)

This word is a combination of two Greek words: 'Gastir which means 'stomach' and podi which means 'foot'. So Gastropods can be called 'stomach-footed' mollusks. As the name suggests these have a well developed muscular foot, sometimes quite large that is used for crawling. There is a head with well developed eyes and tentacles. Except the slugs all gastropods produce an outer shell which is spirally wound. A kind of spiral coiling is seen as the animal grows. The immature organism is bilaterally symmetrical but acquires an asymmetrical shape as it matures into an adult. The coiling of the visceral mass around the central axis occurs through a process called torsion. This class with 17 orders and many families contain animals whose shell shape and structure also help to identify them. A Cone shell is easily distinguished from a top shell or a cowry. They live on land, freshwater and marine environment.

These are animals with a well developed head bearing eyes and tentacles, The foot is usually large and muscular and is especially useful for crawling. Most of the gastropods produce an outer shell which is spirally wound, but this class also includes the slugs which have no shell. A unique characteristic of the gastropods is a process called torsion, in which the immature bilaterally symmetrical animal twists itself into an asymmetrical shape, becoming spirally coiled.

The GASTROPODA is by far the most successful class of all the molluscs, with at least 60,000 living species. Gastropods generally have a single-valved shell, which is usually spiraled; however, this is absent in the slugs and semi-slugs. They have a head with cephalic tentacles and a well-developed foot used in crawling. Some gastropods have lungs for respiration, others gills

Early in their larval stage of development, the visceral mass and mantle cavity rotate up to 180 degrees counterclockwise; in a process know as torsion. (This does not occur in some of the slugs though.) This brings their organs from a posterior position to an anterior position behind their head. In most cases, the soft animal is able to retract into their shells for protection. Some gastropods also have an operculum (trap door) connected to their foot that they can pull in after their soft body parts sealing off their shell from the environmental hazards or as protection from enemies. Gastropods are very successful in marine, freshwater and terrestrial habitats. Their size is from 0.5 to 750 millimeters. (*i.e.*, up to 2 1/2 feet long)

(2) Bivalvia (Bivalves), also called Pelecypoda

(Two shelled like clams, mussels etc). The names mean "two-valved" and "hatchet-footed" respectively. An older name for the class was LAMELLI-BRANCHIATA.

These animals produce two shells (valves) that are connected by a hinge. The animals have no head but with a foot that often extends between the two valves. Common names for some of the bivalves are clams, scallops, and oysters.

(3) Schaphopoda (Scaphopods)

The name means "boat-footed"

Animals without eyes, gills, or tentacles. They are enclosed in tusk-shaped shells, open at each end, which gives them their common name, tusk shells.

(4) Aplacophora (Solenogasters)

The name means "without plates" (without shell). (solenogasters)

Worm-like animals having no shell. The animal has no head, mantle or nephridia (excretory organs). The foot has been reduced to a short ridge down the middle of the animal. Instead of a shell the animal is covered with tiny calcareous spicules.

(5) Polyplacophora (Chitons)

The name means "bearing many plates". (or Amphineura as it was earlier called-Chitons)

These are bilaterally symmetrical, flat molluscs with a shell consisting of eight overlapping plates called valves, which are attached underneath and around the edges to a tough, flexible part of the mantle called the girdle. The head has no eyes or tentacles. The animal has a radula designed for heavy-duty rasping.

(6) Monoplacophora (Gastroverms and segmented limpets)

The name means "bearing one plate" (bearing one shell). (Segmented limpets).

Bilaterally symmetrical cap-shaped shells. The animal is without torsion, but with head, foot, and mantle, and possessing a radula. The muscles, external gills, nerve branches, and nephridia are paired.

(7) Cephalopoda (Cephalopods)

The name means "head-footed".(nautilus, squids, Cuttelfish, octopus etc)

This class is characterized by animals with distinct heads and large, complex eyes. The foot has been altered into a set of arms that circle the mouth and parrot-like jaws ("beak"). Except for the chambered nautilus, the molluscs of this class do not have an external shell. Common names for cephalopods are octopus, squid, cuttlefish, and nautilus (**Plate 50-54**).

Plate 50: Cephalopods.

Plate 51: Squid.

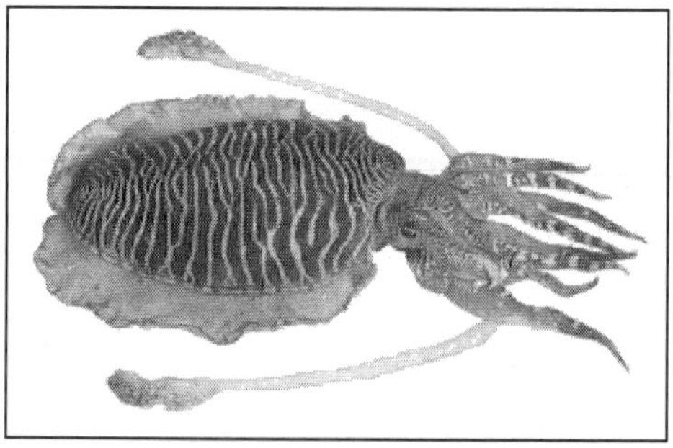

Plate 52: Cuttlefish.

Plate 53: Octopus.

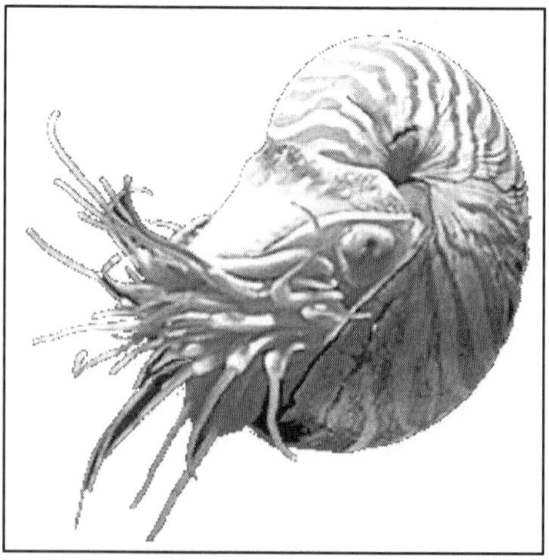

Plate 54: Nautilus.

Complete classification of this animal is therefore:

Kingdom: Animalia

Phylum: Mollusca

Class: Gastropoda

Order: Prosobrachia

Family: Strombidae

Genus: Strombus

Species: marginatus

Subspecies: septimus

Snail

Scientific Classification

Kingdom: Animalia

Phylum: Mollusca

Class: Gastropoda

Overview

Snails that respire using a lung belong to the group Pulmonata, while those with gills form a polyphyletic group; in other words, snails with gills form a number of taxonomic groups that are not necessarily more closely related to each other than they are related to some other groups. Both snails that have lungs and snails that have gills have diversified so widely over geological time that a few species with gills can be found on land and numerous species with lungs can be found in freshwater. Even a few marine species have lungs.

Snails can be found in a very wide range of environments, including ditches,deserts, and the abyssal depths of the sea. Although land snails may be more familiar to people, marine snails constitute the majority of snail species, and have much greater diversity and a greater biomass. Numerous kinds of snail can also be found in fresh water.

Most snails have thousands of microscopic tooth-like structures located on a ribbon-like tongue called a radula. The radula works like a file, ripping food into small pieces. Many snails are herbivorous, eating plants or rasping algae from surfaces with their radulae, though a few land species and many marine species are omnivores or predatory carnivores.

Several species of the genus *Achatina* and related genera are known as **giant African land snails**; some grow to 15 in (38 cm) from snout to tail, and weigh 1 kg (2 lb). The largest living species of sea snail is *Syrinx aruanus*; its shell can measure up to 90 cm (35 in) in length, and the whole animal with the shell can weigh up to 18 kg (40 lb) **(Plates 55-57).**

Plate 55: *Helix aspersa* Garden Snail .

Plate 56: *Helix pomatia* Land Snail.

Plate 57: French Cooked Snails.

Human Relevance

Land snails are known as an agricultural and garden pest but some species are an edible delicacy and occasionally household pets.

In Agriculture

There are a variety of snail-control measures that gardeners and farmers use in an attempt to reduce damage to valuable plants. Traditional pesticides are still used, as are many less toxic control options such as concentrated garlic orwormwood solutions. Copper metal is also a snail repellent, and thus a copper band around the trunk of a tree will prevent snails from climbing up and reaching the foliage and fruit. Placing crushed egg shells on the soil around garden plants can also deter snails from coming to the plants.

The decollate snail (*Rumina decollata*) capture and eat garden snails, and because of this it has sometimes been introduced as a biological pest control agent. However, this is not without problems, as the decollate snail is just as likely to attack and devour other gastropods that may represent a valuable part of the native fauna of the region.

As Food

In French cuisine, edible snails are served for instance in *Escargot à la Bourguignonne*. The practice of rearing snails for food is known as heliciculture. For purposes of cultivation, the snails are kept in a dark place in a wired cage with dry straw or dry wood. Coppiced wine-grape vines are often used for this purpose. During the rainy period the snails come out of hibernation and release most of their mucus onto the dry wood/straw. The snails are then prepared for cooking. Their texture when cooked is slightly chewy.

As well as being relished as gourmet food, several species of land snails provide an easily harvested source of protein to many people in poor communities around the world. Many land snails are valuable because they can feed on a wide range of agricultural wastes, such as shed leaves in banana plantations. In some countries, giant African land snails are produced commercially for food.

Land snails, freshwater snails and sea snails are all eaten in a number of countries (principally Spain, Philippines,Morocco, Nigeria, Algeria, Cameroon, France, Italy, Portugal, Greece, Bulgaria, Belgium, Vietnam, Laos, Cambodia, Cyprus, Ghana, Malta, Terai of Nepal, south western China, Northeast India states such as Manipur, Tripura and parts of the United States). In certain parts of the world, snails are fried. For example, in Indonesia, they are fried as satay, a dish known as *sate kakul*. The eggs of certain snail species are eaten in a fashion similar to the way caviar is eaten.

In Bulgaria snails are traditionally cooked in an oven with rice or fried in a pan with vegetable oil and red paprika powder. Before they are used for those dishes however, they are thoroughly boiled in hot water (for up to 90 minutes) and manually extracted from their shells. The two species most commonly used for food in the country are ***Helix lucorum*** and ***Helix pomatia***.

1. The Giant African Land Snail (*Achatina fulica*)

Classification

Kingdom: Animalia

Phylum: Mollusca

Class: Gastropoda

Order: Achatinoidea

Family: Achatinidae

Genus: Achatina

Scientific Name: ***Achatina Fulica hamillei*** **Petit**

Type: Mollusc

Diet: Herbivore

Size: 15-30cm (6-11.8in)

Weight: 250-450g (8.8-15.9oz)

Top Speed: 0.003km/h

Life Span: 3-10 years

Lifestyle: Solitary

The giant African land snail is the largest species of snail found on land and generally grow to around 20 cm in length. The giant African land snail is native to the forest areas of East Africa but has been introduced into Asia, the Caribbean and a number of islands in both the Pacific and the Indian oceans **(Plates 58-60).**

The giant African land snail is generally seen as a pest as these snails will eat almost anything vegetarian that they can find and have proven to be quite destructive when around crops and wild flowers. Giant African land snails are also known to carry parasites and are illegal to keep as pets in some countries such as the USA.

The giant African land snail is native to humid, forest areas but can today be found in agricultural areas, coast land, natural forest, planted forests, shrublands, urban areas, and wetlands. The giant African land snail is seen to be highly invasive species and large colonies of land snails can be formed from just one individual.

Plates 58–60: Giant African Land Snail (*Achatina fulica*).

Giant African land snails have both male and female reproductive organs. Although giant African land snails primarily mate with one another, in more isolated regions the giant African land snail is capable of reproducing by itself.

Giant African land snail lay around 6 clutches of eggs every year, laying an average of 200 eggs per clutch. Around 90 per cent of snail hatchings survive meaning that a snail free area can quickly become infested.

Giant African land snails are active during the night and spend the daytime hours safely buried underground. Giant African land snails reach their adult size by the time they are 6 months old and although their growth rate slows at this point, giant African land snails never stop growing. Most giant African land snail reach between 5 and 6 years of age but some giant African land snail individuals have been known to be more than 10 years old.

During periods of extreme drought, the giant African land snail goes into aestivation (summer sleep). The giant African land snail seals itself inside it's shell to retain water and giant African land snails do this about 3 times a year, depending on the areas in which they inhabit.

Achatina fulica is a species of land snail in the family **Achatinidae**, known commonly as the **giant African snail** or **giant African land snail**. Outside of its native range this snail thrives in many types of habitat in areas with mild climates. It feeds voraciously and is a vector for plant pathogens, causing severe damage to agricultural crops and native plants. It competes with native snail taxa, is a nuisance pest of urban areas, and spreads human disease. This snail is listed as one of the top 100 invasive species in the world

Distribution

The species is native to East Africa, but it has been widely introduced to other parts of the world through the pet trade, as a food resource, and by accidental introduction.

This species has been found in China since 1931, and its initial point of distribution in China was Xiamen. The snail has also been established in the Pratas Islands, of Taiwan, throughout **India,** the Pacific, Indian Ocean islands, and the West Indies. In the United States, it has become established in Hawaii and eradication is underway in Florida.

The species has recently been observed in Bhutan (Gyelposhing, Mongar), where it is an invasive species. It has begun to attack agricultural fields and flower gardens. It is believed there that dogs have died as a result of consuming the snail.

Description

The adult snail is around 7 centimetres (2.8 in) in height and 20 centimetres (7.9 in) or more in length. The shell has a conical shape, being about twice as high as it is broad. Either clockwise (dextral) or counter-clockwise (sinistral) directions can be observed in the coiling of the shell, although the right-handed (dextral) cone is the more common. Shell colouration is highly variable, and dependent on diet. Typically, brown is the predominant colour and the shell is banded. The shell is particularly tough and has the highest heavy metal content of any snail species.

Habitat

The giant African snail is native to East Africa, and can be traced back to Kenya and Tanzania. It is a highly invasive species, and colonies can be formed from a single **gravid** individual. In many places, release into the wild is illegal. Nonetheless, the species has established itself in some temperate climates and its habitat now includes most regions of the humid tropics, including many Pacific islands, southern and eastern Asia, and the Caribbean. The giant snail can now be found in agricultural areas, coastland, natural forest, planted forests, riparian zones, scrub and shrublands, urban areas, and wetlands.

Feeding

The giant African snail is a macrophytophagous **herbivore;** it eats a wide range of plant material, fruit, and vegetables. It will sometimes eat sand, very small stones, bones from carcasses and even concrete as calcium sources for its shell. In rare instances the snails will consume each other.

In captivity, this species can be fed on grain products such as bread, digestive biscuits, and chicken feed. It can be supplemented with calcium for shell development. It requires about 18.28 per cent of crude protein in its diet for optimal growth.

Life Cycle

This species is a simultaneous **hermaphrodite;** each individual has both **testes** and **ovaries** and is capable of producing both sperm and ova. Instances of self-fertilization are rare, occurring only in small populations. Although both snails in a mating pair can simultaneously transfer gametes to each other (bilateral mating), this is dependent on the size difference between the partners. Snails of similar size will reproduce in this way. Two snails of differing sizes will mate unilaterally (one way), with the larger individual acting as a female. This is due to the comparative resource investment associated with the different genders.

Like other land snails, these have intriguing mating behaviour, including petting their heads and front parts against each other. Courtship can last up to half an hour and the actual transfer of gametes can last for two hours. Transferred sperm can be stored within the body for up to two years. The number of eggs per clutch averages around 200. A snail may lay five to six clutches per year with a hatching viability of about 90 per cent.

Adult size is reached in about six months, after which growth slows, but does not cease until death. Life expectancy is commonly five or six years in captivity, but the snails can live for up to ten years. They are **active at night** and spend the day buried underground. The giant African snail is capable of aestivating for up to three years in times of extreme drought, sealing itself into its shell by secretion of a calcerous compound that dries on contact with the air **(Plates 61-63).**

Plate 61: Fresh Eggs.

Plate 62: Juvenile Snail.

Plate 63: Hatching from Eggs.

Giant African Land Snail (*Achatina fulica*).

Parasites

Parasites of *Achatina fulica* include:

- *Aelurostrongylus abstrusus*
- *Angiostrongylus cantonensis* – causes eosinophilic meningoencephalitis
- *Angiostrongylus costaricensis* – causes abdominal angiostrongyliasis
- *Schistosoma mansoni* – causes schistosomiasis, detected in faeces
- *Trichuris* spp. – detected in faeces
- *Hymenolepis* spp. – detected in faeces
- *Strongyloides* spp. – detected in faeces and in mucous secretion

As an Invasive Species

In many places this snail is a **pest** of agriculture and households with the ability to transmit both human and plant pathogens. Suggested preventive measures include strict quarantine to prevent introduction and further spread. This snail has been given top national quarantine significance in the United States. In the past, quarantine officials have been able to successfully intercept and eradicate incipient invasions on the mainland USA.

In the wild, this species often harbors the parasitic nematode *Angiostrongylus cantonensis,* which can cause a very serious meningitis in humans. Human cases of this meningitis usually result from a person having eaten the raw or undercooked snail, but even handling live wild snails of this species can infect a person with the nematode and thus cause a life-threatening infection.

In some regions, an effort has been made to promote use of the giant African snail as a food resource in order to reduce its populations. However, promoting a pest in this way is a controversial measure, because it may encourage the further deliberate spread of the snails.

Plate 64: Adult Shell.

Plate 65: Adult and Juvenile.

Plate 66: Feeding on a Leaf.

Plate 67: Apertural View of the Shell.

Giant African Land Snail (*Achatina fulica*)

One particularly catastrophic attempt to biologically control this species occurred on South Pacific Islands. Colonies of *A. fulica* were introduced as a food reserve for the American military during World War II and they escaped. A carnivorous species (Florida rosy wolfsnail, *Euglandina rosea*) was later introduced by the United States government, in an attempt to control *A. fulica* but the rosy wolf snail instead heavily preyed upon the native *Partula*, causing the extinction of most *Partula* species within a decade **(Plates 64-67).**

Human Use

Achatina fulica are used by some practitioners of **Candomblé** for religious purposes in Brazil as an offering to the deity Oxalá. The snails substitute for a closely related species, the African giant snail (*Archachatina marginata*) normally offered in Nigeria. The two species share a common name (*Ìgbín*, also known as *ibi* or *boi-de-oxalá* in Brazil), and are similar enough in appearance to satisfy religious authorities. They are also edible if cooked properly.

2. Helix Species

Helix aspersa - UP, Bihar, Maharashtra, Odissa and Gujrat

H nemoralis - UP, Bihar, Maharashtra, Odissa and Gujrat

H hortensis- UP, Bihar, Maharashtra, Odissa and Gujrat

Viviparus variatus- South India

Xestina ligulata- Travancore, Cochin, Annamalai

Ariophanta bisarialis- Travancore, Cochin, Annamalai

Helix aspersa (*Cornu aspersum*)

Classification

Kingdom: Animalia

Class: Gastropoda

Superfamily: Helicoidea

Family: Helicidae

Genus: *Cornu*

Species: *C. aspersu*

The adult bears a hard, thin **calcareous** shell 25–40 mm in diameter and 25–35 mm high, with four or five **whorls**. The shell is variable in color and shade but generally is dark brown, brownish golden, or chestnut with yellow stripes, flecks, or streaks (characteristically interrupted brown colour bands).The aperture is large and characteristically oblique, its margin in adults is white and reflected **(Plate 68).**

The body is soft and slimy, brownish-grey, and the animal retracts itself entirely into the shell when inactive or threatened. When injured or badly irritated the animal produces a defensive froth of mucus that might repel some enemies or overwhelm aggressive small ants or the like. It has no **operculum**; during dry or cold weather it

Plate 68: *Cornu aspersum* in Warm Regions.

seals the aperture of the shell with a thin membrane of dried mucus; the term for such a membrane is **epiphragm**. The epiphragm helps the snail retain moisture and protects it from small predators such as some ants.

The snail's **quiescent** periods during heat and drought are known as **aestivation;** its quiescence during winter is known as **overwintering**. When overwintering, *Cornu aspersum* avoids the formation of ice in its tissues by altering the osmotic components of its blood (or haemolymph); this permits it to survive temperatures as low as -5 °C (23 °F). During aestivation, the **mantle** collar has the ability to change its permeability to water. The snail also has an **osmoregulatory** mechanism that prevents excessive absorption of water during hibernation. These mechanisms allow *Cornu aspersum* to avoid either fatal desiccation or hydration during months of either kind of quiescence.

During times of activity the snail's head and "foot" or "**belly" emerge.** The head bears four **tentacles;** the upper two are larger and bear eye-like light sensors, and the lower two are tactile and olfactory sense organs. The snail extends the tentacles by internal pressure of body fluids, and retracts all four tentacles into the head by **invagination** when threatened or otherwise retreating into its shell. The mouth is located beneath the tentacles, and contains a chitinous radula with which the snail scrapes and manipulates food particles.

Life Cycle

Like other Pulmonata, *Cornu aspersum* is a **hermaphrodite**, producing both male and female **gametes.** Reproduction is usually **sexual**, although **self-fertilisation** sometimes occurs. During a mating session of several hours, two snails exchange sperm. *Cornu aspersum* is one of the species that uses **love darts** during mating.

About two weeks after fertilisation, the snail lays a batch of about 80 spherical pearly-white eggs into crevices in the topsoil or sheltered under stones or the like. In a year it may lay six batches or so. The size of the egg is 4 mm.

The young snails take one to two years to reach maturity. In some regions snail farms produce them commercially.

Ecology

Cornu aspersum is a primarily a **herbivore** with a wide range of host plants. It feeds on numerous types of **fruit trees, vegetable** crops, garden **flowers,** and **cereals**. It also is an **omnivorous** scavenger that feeds on rotting plant material and on occasion will scavenge animal matter, such as crushed snails and worms. In turn it is a food source for many other animals, including small mammals, many bird species, lizards, frogs, centipedes, predatory insects such as glowworms in the family Lampyridae, and predatory terrestrial snails. The species may on occasion be of use as an indicator of environmental pollution, because it deposits heavy metals, such as lead in its shell.

Parasites

The nematode *Nemhelix bakeri* (family Cosmocercidae)

Behaviour

The snail secretes thixotropic adhesive **mucus** that permits locomotion by rhythmic waves of contraction passing forward within its muscular **"foot".** Starting from the rear, the contraction of the longitudinal muscle fibres above a small area of the film of mucus causes shear that liquefies the mucus, permitting the tip of the tail to move forward. The contracted muscle relaxes while its immediately **anteriad** transverse band of longitudinal fibres contract in their turn, repeating the process, which continues forward until it reaches the head. At that point the whole animal has moved forward by the length of the contraction of one of the bands of contraction. However, depending on the length of the animal, several bands of contraction can be in progress simultaneously, so that the resultant speed amounts to the speed imparted by a single wave, multiplied by the number of individual waves passing along simultaneously.

A separate type of wave motion that may be visible from the side enables the snail to conserve mucus when moving over a dry surface. It lifts its belly skin clear of the ground in arches, contacting only one to two thirds of the area it passes over. With suitable lighting the lifting may be seen from the side as illustrated, and the percentage of saving of mucus may be estimated from the area of wet mucus trail dabs that it leaves behind. This type of wave passes backwards at the speed of the snail's forward motion, therefore having a zero velocity with respect to the ground.

The snail moves at a top speed of 1.3 centimetres per second (47 meters per hour or ~50 yards per hour), and has a strong **homing** instinct, readily returning to a regular **hibernation** site.

In spite of its apparent slowness and limitations, the snail exploits the special nature of its mucus to achieve some startling feats. It can go up a slope at any angle, including upside down, resist being pulled off a firm surface with an adhesive strength several times its own weight, rest on a surface at any angle without any expenditure of energy, or, notoriously, climb a needle-like stem or pass over the edge of razor blade without harm, relying on the firmness of its mucus film in its shear-resistant phase.

3. *Bensonia monticola*

J and K, Himanchal and Uttarakhand

4. *Macroclamis indica* (Garden snail)

Assam, Bihar, W. Bengal, Odissa, T.N. and Delhi.

5. *Opis gracilis* (Green house)

Assam, Manipur, W. Bengal, Bihar, Kerala, T.N., Maharashtra, Rajasthan, U.P. and Delhi.

Slugs

Classification

Kingdom: Animalia

Phylum: Mollusca

Class: Gastropoda

Superfamily: Veronicelloidea

Family: Veronicellidae

Genus: ***Laevicaulis***

Species: ***L. alte***

Gastropods that lacks a conspicuous shell are commonly called **slugs** rather than snails. Some species of slug have a red shell, some have only an internal **vestige** that serves mainly as a calcium repository, and others have no shell at all. Other than that there is little morphological difference between slugs and snails. There are however important differences in habitats and behavior.

A shell-less animal is much more maneuverable and compressible, so even quite large land slugs can take advantage of habitats or retreats with very little space, retreats that would be inaccessible to a similar-sized snail. Slugs squeeze themselves into confined spaces such as under loose bark on trees or under stone slabs, logs or wooden boards lying on the ground. In such retreats they are in less danger from either predators or desiccation, and often those also are suitable places for laying their eggs.

Slugs as a group are far from **monophyletic;** biologically speaking "slug" is a term of convenience with little taxonomic significance. The reduction or loss of the shell has evolved many times independently within several very different lineages of gastropods. The various taxa of land and sea gastropods with slug morphology occur within numerous higher taxonomic groups of shelled species; such independent slug taxa are not in general closely related to one another. ***Laevicaulis alte***, is a species of tropical land slug, a terrestrial pulmonate gastropod mollusk in the family Veronicellidae, the leatherleaf slugs.

Plate 69: *Laevicaulis alte.*

Plate 70: Ventral View of Foot of Juvenile.

Plate 71: Contracted Individual.

Description

Laevicaulis alte is a round, dark-coloured slug with no shell, 7 or 8 cm long. Its skin is slightly tuberculated. The central keel is beige in colour. **(Plate 69-71)**

This slug has a unique, very narrow foot; juvenile specimens have a foot 1 mm wide and adult specimens have a foot that is only 4 or 5 mm wide.

The **tentacles** are small, 2 or 3 mm long, and they are only rarely extended beyond the edge of the **mantle**.

Parasites

This slug is an intermediate host for *Angiostrongylus cantonensis*, the rat lungworm, a round worm, the most common cause of eosinophilic meningoencephalitis.

Predators

This slug is eaten by the frog *Rana tigrina*.

Distribution

This species is probably indigenous to Africa (western Africa and eastern Africa).Tanzania

It has been introduced and become an invasive species in the following areas:

- Southern Asia
- Pratas Islands, Taiwan
- United States (Hawaii)
- islands in the Indian Ocean
- Australia (since 1889)
- Samoa

Habitat

This species lives in dry areas, mostly at lower altitudes.

Feeding Habits

What this slug eats in the wild is not known. In captivity this slug eat:

- Dandelion leaves
- Tomatoes - it nibbles the skin first and then the pulp
- Cucumber
- Apples
- Spinach
- Grapes

Life Cycle

This slug hatches from eggs. This species has several adaptations for living in dry conditions: a rounded shape with as small as possible surface area, and a narrow foot to reduce evaporation.

Juvenile specimens search for food nearly always at night, and stay buried in the soil during the day. Larger specimens are active during the day sometimes. This slug can grow up from 0.5 cm to approximately 4 cm in length in 7 months.

Mariaella dussumieri

Classification

Kingdom: Animalia

Phylum: Mollusca

Class: Gastropoda

Superfamily: Helicarionoidea

Family: Ariophantidae

Genus: ***Mariaella***

Species: ***M. dussumieri***

Mariaella dussumieri is a species of air-breathing land slug, a terrestrial pulmonate gastropod mollusk in the family Ariophantidae, a slug from the Western Ghats of India. The species is tentatively identified as *Mariaella dussumieri*. It is the type species of the genus *Mariaella*. Its specific name is in honor of French merchant Jean-Jacques Dussumier.

Distribution

This species lives in India. It is endemic to Western Ghats. In the Western Ghats it occurs as far north as the Kadur district of Mysore and perhaps to Mahableshwar and also in hill-tracts of Ceylon.

This species has not yet become established in the USA, but it is considered to represent a potentially serious threat as a pest, an invasive species which could negatively affect agriculture, natural ecosystems, human health or commerce. Therefore it has been suggested that this species be given top national quarantine significance in the USA.

Shell Description

The internal shell, in general thin, slightly convex, ovate and white. The

Plate 72: ***Mariaella dussumieri*** **Adult.**

Plate 73: ***Mariaella dussumieri*** **Mating.**

apex of the shell is forming a small point at the right side and near the posterior end. The shell is not involute.

Anatomy

The color of the animal is yellowish brown or olivaceous, sometimes almost black, and generally mottled with dark blotches. The mantle has two narrow raised ridges on the shell-lobes, one running from the little shell-aperture round the left margin of the shell, the other towards the respiratory orifice on the right margin. The size appears to vary up to about 8 inches (200 mm) in length **(Plates 72-73).**

Chapter 5
Rodents: Field and Storage Species

General Information

Rodents (from Latin *rodere,* "to gnaw") are mammals of the order **Rodentia**, which are characterized by a single pair of unremittingly growing **incisors** in each of the upper and lower **jaws**. About forty per cent of all mammal species are rodents; they are found in vast numbers on all continents except Antarctica. They are the most diversified mammalian order and live in a variety of terrestrial habitats, including human-made environments. There are species that are **arboreal, fossorial** (burrowing), and semiaquatic. Well-known rodents include **mice, rats, squirrels, prairie dogs, porcupines, beavers, guinea pigs, hamsters**, and **capybaras**. Other animals such as rabbits, hares and pikas were once included with them, but are now considered to be in a separate order, **Lagomorpha**.

Most rodents are small animals with robust bodies, short limbs and long tails. They use their sharp incisors to gnaw food, excavate burrows and defend themselves. Most eat seeds or other plant material, but some have more varied diets. They tend to be social animals and many species live in societies with complex ways of communicating with each other. Mating among rodents can vary from **monogamy,** to **polygyny,** to **promiscuity**. Many have litters of underdeveloped, **altricial** young, while others have **precocial** young that are relatively well developed at birth.

The rodent fossil record dates back to the Paleocene on the supercontinent of Laurasia. They greatly diversified in the Eocene, as they spread across continents, sometimes even finding means to cross oceans. Rodents reached both South America and Madagascar from Africa, and were the only terrestrial placental mammals to reach and colonize Australia.

Rodents have been used as food, for clothing, as pets and as laboratory animals in research. Some species, in particular the brown rat, the black rat, and the house mouse are serious pests, eating and spoiling food stored by humans, and spreading diseases. Accidentally introduced species of rodents are often considered to be invasive, and have caused the extinction of numerous species, such as island birds, previously isolated from land-based predators.

Rats are various medium-sized, long-tailed rodents of the **superfamily Muroidea**. **"True rats"** are members of the genus *Rattus*, the most important of which to humans are the **black rat,** *Rattus rattus*, and the **brown rat**, *Rattus norvegicus*. Many members of other rodent genera and families are also referred to as rats, and share many characteristics with true rats.

Rats are typically distinguished from **mice** by their size. Generally, **when someone discovers a large muroid rodent, its common name includes the term *rat*, while if it is smaller; the name includes the term *mouse*.** The muroid family is broad and complex, and the common terms ***rat*** and ***mouse*** are not taxonomically specific. Scientifically, the terms are not confined to members of the *Rattus* and *Mus* genera, for example, the pack rat and cotton mouse.

1. Black Rat, Roof Rat, Ship Rat, House Rat (*Rattus rattus*)

Classification

Kingdom: Animalia

Phylum: Chordata

Class: Mammalia

Order: Rodentia

Family: Muridae

Genus: *Rattus*

Species: *R. rattus*

Binomial name: *Rattus rattus*Thomas, 1910

Habitat

Found in all habitats except cold deserts. *Rattus rattus* is most often found in large numbers in coastal areas because of the way the species is spread through human sea faring. It is generally found in any area that can support its mainly vegetarian diet. Because *R. rattus* is an agile climber, it often lives in high places, such as top floors of buildings in populated areas or trees in forested areas. Even though it can be found near water, this species rarely swims and unlike its close relatives, rarely finds a home in sewers or in aquatic areas. Although it was formerly common in towns and farms of temperate regions, it has been largely driven out by the more aggressive Norway rat as well as killed off by increasing chemical pest control programs. Data have shown that *R. rattus* can reach elevations up to 250 m above sea level.

Plate 74: House Rat (*Rattus rattus*).

Non-Migrant: Yes. At least some populations of this species do not make significant seasonal migrations. Juvenile dispersal is not considered a migration.

Locally Migrant: No populations of this species make local extended movements (generally less than 200 km) at particular times of the year (*e.g.*, to breeding or wintering grounds, to hibernation sites).

Associations

They compete with other species of rodents, such as *Rattus norvegicus. Rattus rattus* is a disease vector, responsible for **bubonic plague** outbreaks and other diseases. This cosmopolitan species hosts a wide variety of internal and external parasites, up to 18 species of gastrointestinal helminths in some areas.

Distinguishing Features

Body size is about 8cm in length. The upper body colouring is brown, black or charcoal grey, with a cream or white underbelly. The tail is scaly and long relative to its body size. Front teeth are chisel shaped.

Morphology

Rattus rattus is a medium sized rat with relatively large ears and a tail that is nearly always longer than the body. Individuals weigh between 70 and 300 g, and are between 16 and 22 cm in head and body length and a tail length of 19 cm or longer. Males are longer and heavier than are females. Many members of the species are black in color with a lighter colored ventral belly. The species is often divided into subspecies based upon color patterns which can occur in any combination of black, white, grey, and agouti. The skull and nasal bones are relatively narrow. One of the main ways to differentiate between *R. rattus* and *R. norvegicus* is that *R. rattus* has a finer covering of hair, a lighter skull, and a slightly differently shaped upper first molar. **(Plate 74)**

Similar Species

The Australian native Bush Rat (*Rattus fuscipes)* and the Agile Antechinus (*Antechinus agilis*).

Distribution

This species is found widely across Australia, especially in the coastal regions in human modified areas. It is also found commonly in and around human settlements where food and shelter is readily available in most countries of the world.

Associated Vegetation Community

In their natural environment Black Rats are common to woodlands, forests and grasslands. They are typically associated with human settlements, including cities, rural villages and farms.

Limiting Resources

Black Rats are very adaptable and can live in a variety of places such as roofs, cavity walls,trees, scrapes or burrows around farms. A close distance to humans usually ensures the food and water requirements of this species are met.

Breeding

A mature female rat can give birth to about 20 young in a year (4 to 6 at a time). Young are born in nests made from paper, rubbish, plastics, grass or other material. They become completely independent and able to reproduce after three months. Outdoor populations in the natural environment tend to peak in summer or early autumn, while indoor populations remain constant throughout the year.

Behaviour

Black Rats are very good swimmers and climbers, meaning they can access most areas to feed, shelter and breed. Usually they move about 50m from their nest site. They will climb trees to steal eggs out of birds' nests or to eat fruit. They are known to cause trouble by gnawing at pipes and electrical cords. Rats have very poor vision, relying on their keen sense of smell, acute hearing and their whiskers and sense of touch. They spend time around water bodies and beaches and are the most commonly seen rodent in public parks and beaches.

Food Species

In the wild, Black Rats eat a range of fresh plant material, nuts and seeds, fruit and vegetables and tree bark. In human environments they are known to eat almost anything.

Predators

Feral or domestic Cats (*Felis catus)* and Dogs (*Canis familiaris*), snakes, birds of prey, and other rats all attack and kill Black Rats

Interesting Fact

If the mother rat has become wary of rodenticides or traps, many of her young will learn to avoid them. This learning experience can make management difficult in sites where long-term rodent suppression programs have been unsuccessful in the past.

Biology

The black rat is **nocturnal,** although it may become more active in the day in undisturbed areas. It is an **omnivore,** but tends to prefer plant matter such as fruits and seeds, although it will also feed on insects, carrion, refuse and faeces. On Lundy Island these rats feed on crabs along the shore. This rat lives in groups called 'packs', consisting of several males and two or more dominant females. They are skilled climbers and can also swim well. Nests are constructed from grass and twigs, often in roof spaces, a habit which earned the species the further common name of 'roof rat'. Breeding takes place between March and November; 3 to 5 litters can be produced in a year, each litter containing 1-16 young (although the average is 7). A single female can therefore produce a huge number of offspring; 56 young were recorded on a London ship for a single female. At 12-16 weeks of age, females are capable of breeding; they are also able to conceive whilst still suckling the previous litter, which further maximises their reproductive capability.Maximum lifespan in the wild is less than 18 months; populations have very high mortality rates, mainly as a result of widespread pest control measures. The black rat is a notorious pest, and was the host of the fleas that carried bubonic plague. It also carries a host of other diseases and is damaging to property and food stores.

Range mass: 70 to 300 g.

Average mass: 200 g.

Range length: 16 to 22 cm.

Other Physical Features: endothermic; homoiothermic; bilateral symmetry

Sexual Dimorphism: male larger

Average basal metabolic rate: 0.77 W.

Length: 46 cm

Weight: 350 grams

Reproduction

Social groups of *R. rattus* are often formed of multiple males and multiple females. One male is dominant and a linear male hierarchy may form. Two to three females are often dominant to all other group members except the dominant male. Females are generally more aggressive than males. The species is polygynous, and generally, the dominant male is the most successful breeder. Territories and mates are defended through aggressive behavior. If environmental conditions allow it, successful breeding may occur all year.

Mating System: polygynous

Rattus rattus is able to breed throughout the year if conditions allow. The peak breeding seasons are summer and autumn. Females can produce up to 5 litters in one year. The gestation period ranges between 21 and 29 days, and young rats are able to reproduce within 3 to 5 months of their birth. Neonates are altricial, like most rodents, and their eyes do not open until 15 days of age. Young remain hairless for much of their nursing period. Weaning and independence from the mother occur at about 3 to 4 weeks of age.

Breeding interval: *R. rattus* breeds year-round producing up to five litters in that time.

Breeding season: *R. rattus* mates throughout the year if environmental conditions permit, however peak times are summer and autumn seasons.

Range number of offspring: 6 to 12.

Average number of offspring: 8.

Range gestation period: 21 to 29 days.

Range weaning age: 3 to 4 weeks.

Range time to independence: 3 to 4 weeks.

Range age at sexual or reproductive maturity (female): 3 to 5 months.

Range age at sexual or reproductive maturity (male): 3 to 5 months.

Key Reproductive Features: iteroparous; year-round breeding; gonochoric/ gonochoristic/dioecious (sexes separate); sexual; fertilization; viviparous

Average birth mass: 4.55 g.

Average number of offspring: 7.3.

Because male members of *R. rattus* copulate with one female and then move on to the next, they don't contribute much to the care of the young. The young remain relatively helpless for about 2 weeks, until they begin to grow a pelage, their eyes open, and they are able to move around more. Weaning is accompanied by increased independence from the mother. Until these rats reach their full adult size, they stay in the nest built by their mother. Young rats are capable of reproducing by about 3 to 5 months of age.

2. Brown Rat (*Rattus norvegicus*)

Other Names

Common rat, street rat, sewer rat, Hanover rat, Norway rat, brown Norway rat, Norwegian rat, or wharf rat.

Classification

Kingdom: Animalia

Phylum: Chordata

Class: Mammalia

Order: Rodentia

Family: Muridae

Genus: ***Rattus***

Species: ***R. norvegicus***

Binomial name: ***Rattus norvegicus*** **Berkenhout, 1769**

Plate 75: Brown Rat (*Rattus norvegicus*).

Naming and Etymology

Originally called the **"Hanover rat"** by people wishing to link problems in 18th century England with the House of Hanover, it is not known for certain why the brown rat is named *Rattus norvegicus* (Norwegian rat), as it did not originate from Norway. However, the English naturalist John Berkenhout, author of the 1769 book *Outlines of the Natural History of Great Britain,* is most likely responsible for popularizing the misnomer. Berkenhout gave the brown rat the binomial name *Rattus norvegicus,* believing it had migrated to England from Norwegian ships in 1728, although no brown rat had entered Norway at that time.

By the early to middle part of the 19th century, British academics were aware that the brown rat was not native to Norway, hypothesizing (incorrectly) that it may have come from Ireland, Gibraltar or across theEnglish Channel with William the Conqueror. As early as 1850, however, a more correct understanding of the rat's origins was beginning to develop. The British novelist Charles Dickens acknowledged the misnomer in his weekly journal, *All the Year Round,* writing: "Now there is a mystery about the native country of the best known species of rat, the common brown rat. It is frequently called, in books and otherwise, the 'Norway rat', and it is said to have been imported into this country in a ship-load of timber from Norway. Against this hypothesis stands the fact that when the brown rat had become common in this country, it was unknown in Norway, although there was a small animal like a rat, but really a lemming, which made its home there."

Academics began to understand the origins and corrected etymology of the brown rat towards the end of the 19th century, as seen in the 1895 text Natural History by American scholar Alfred Henry Miles:

"The brown rat is the species common in England, and best known throughout the world. It is said to have travelled from Persia to England less than two hundred years ago and to have spread from thence to other countries visited by English ships."

Though the assumptions surrounding this species' origins were not yet entirely accurate, by the 20th century, it was established among naturalists that the brown rat did not originate in Norway, rather the species came from central Asia and (likely) China. Despite this, this species' common name of "Norway rat" is still in use today.

Morphology

One of the largest muroids, it is a brown or grey rodent with a body up to 25 cm (10 in) long, and a similar tail length; the male weighs on average 350 g (12 oz) and the female 250 g (9 oz). Thought to have originated in northern China, this rodent has now spread to all continents except Antarctica, and is the dominant rat in Europe and much of North America–making it by at least this particular definition the most successful mammal on the planet after humans. With rare exceptions, the brown rat lives wherever humans live, particularly in urban areas.

Description

The fur is coarse and usually brown or dark grey, while the underparts are lighter grey or brown. The brown rat is a rather large true murid and can weigh twice as much as a black rat and many times more than a **house mouse**. The length is commonly in the range of 20 to 25 cm (8 to 10 in), with the tail a further 18 to 25 cm (7 to 10 in), thus being roughly the same length as the body. Adult body weight averages 350 g (12 oz) in males and about 250 g (9 oz) in females. Exceptionally large individuals can reportedly reach 900 to 1,000 g (32 to 35 oz) but are not expected outside of domestic specimens. Stories of rats attaining sizes as big as cats are exaggerations, or misidentifications of other rodents, such as the **coypu and musk rat**. In fact it is common for breeding wild brown rats to weigh (sometimes considerably) less than 300 g (11 oz) **(Plates 75-77).**

Brown rats have acute hearing, are sensitive to ultrasound, and possess a very highly developed olfactory sense. Their average heart rate is 300 to 400 beats per

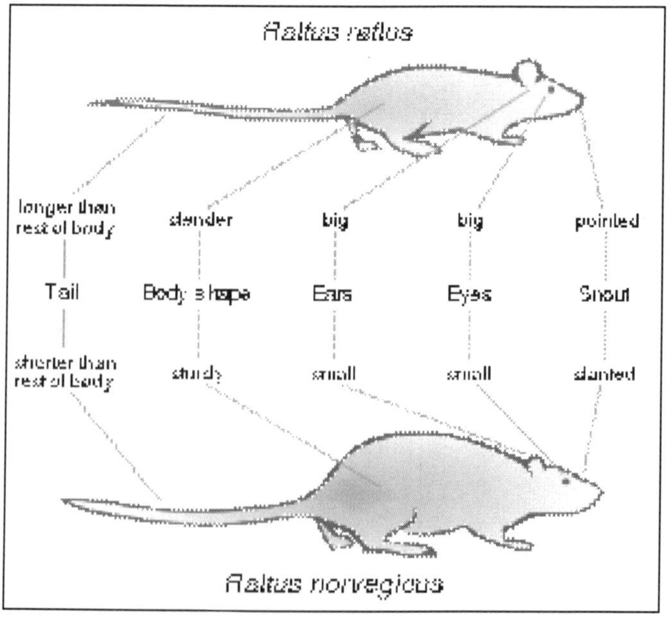

Plate 76: Comparison of the Physique of a Black Rat (*Rattus rattus*) with a Brown Rat (*Rattus norvegicus*).

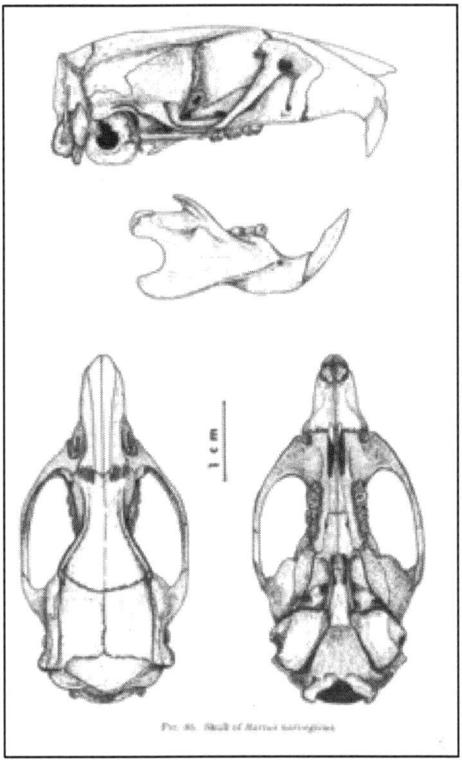

Plate 77: Brown Rat Skull.

minute, with a respiratory rate of around 100 per minute. The vision of a pigmented rat is poor, around 20/600, while a non-pigmented (albino) with no melanin in its eyes has both around 20/1200 vision and a terrible scattering of light within its vision. Brown rats are dichromates which perceive colors rather like a human with **red-green colorblindness**, and their colour saturation may be quite faint. Their blue perception, however, also has UV receptors, allowing them to see ultraviolet lights that some species cannot.

Biology and Behaviour

The brown rat is nocturnal and is a good swimmer, both on the surface and underwater, and has been observed climbing slim round metal poles several feet in order to reach garden bird feeders. Brown rats dig well, and often excavate extensive burrow systems. A 2007 study found brown rats to possess metacognition, a mental ability previously only found in humans and some primates, but further analysis suggested they may have been following simple operant conditioning principles.

Communication

Brown rats are capable of producing ultrasonic vocalizations. As pups, young rats use different types of ultrasonic cries to elicit and direct maternal search behavior, as well as to regulate their mother's movements in the nest. Although pups will produce ultrasounds around any other rats at 7 days old, by 14 days old they significantly reduce ultrasound production around male rats as a defensive response. Adult rats will emit ultrasonic vocalizations in response to predators or perceived danger; the frequency and duration of such cries depends on the sex and reproductive status of the rat. The female rat will also emit ultrasonic vocalizations during mating.

Chirping

Rats may also emit short, high frequency, ultrasonic, socially induced vocalization during rough and tumble play, before receiving morphine, or mating, and when tickled. The vocalization, described as a distinct "chirping", has been likened to laughter, and is interpreted as an expectation of something rewarding. Like most rat vocalizations, the chirping is too high in pitch for humans to hear without special equipment. Bat detectors are often used by pet owners for this purpose.

In clinical studies, the chirping is associated with positive emotional feelings, and social bonding occurs with the tickler, resulting in the rats becoming conditioned to seek the tickling. However, as the rats age, the tendency to chirp appears to decline.

Rat chirp also can be used for mosquito control.

Other ultrasonic vocalizations, including a lower-frequency 'boom' or 'whoom' noise can be produced by bucks in a calm state, when grooming or settling down to sleep.

Audible Communication

Brown rats also produce communicative noises capable of being heard by humans. The most commonly heard in domestic rats is bruxing, or teeth-grinding, which is most usually triggered by happiness, but can also be 'self-comforting' in stressful situations, such as a visit to the vet. The noise is best described as either a quick clicking or 'burring' sound, varying from animal to animal.

In addition, they commonly squeak along a range of tones from high, abrupt pain squeaks to soft, persistent 'singing' sounds during confrontations.

Diet

The brown rat is a true omnivore and will consume almost anything, but cereals form a substantial part of its diet **(Plate 78).**

The most-liked foods of brown rats include scrambled eggs, macaroni and cheese, and cooked corn kernels. According to Schein, the least-liked foods were raw beets, peaches, and raw celery.

Foraging behavior is often population-specific, and varies by environment and food source. Brown rats living near a hatchery in West Virginia catch fingerling fish. Some colonies along the banks of the Po River in Italy will dive for mollusks, a practice demonstrating social learning among members of this species. Rats on the island of Norderoog in the North Sea stalk and kill sparrows and ducks.

Reproduction and Life Sycle

The brown rat can breed throughout the year if conditions are suitable, with a female producing up to five litters a year. The gestation period is only 21 days, and litters can number up to 14, although seven is common. They reach sexual maturity in about five weeks. Under ideal conditions (for the rat), this means the population of females could increase by a factor of three and a half (half a litter of 7) in 8 weeks (5 weeks for sexual maturity and 3 weeks of gestation), corresponding to a population growing by a factor of 10 in just 15 weeks. The maximum life span is up to three years, although most barely manage one. A yearly mortality rate of 95 per cent is estimated, with predators and interspecies conflict as major causes.

When lactating, female rats display a 24-hour rhythm of maternal behavior, and will usually spend more time attending to smaller litters than large ones.

Brown rats live in large, hierarchical groups, either in burrows or subsurface places, such as sewers and cellars. When food is in short supply, the rats lower in social order are the first to die. If a large fraction of a rat population is exterminated, the remaining rats will increase their reproductive rate, and quickly restore the old population level.

Social Behavior

Rats commonly groom each other and sleep together. As with dogs, rats create a social hierarchy, and each rat has its own place in the pack. Rats are said to establish an order of hierarchy, so one rat will be dominant over another one. Groups of rats tend to "play fight", which can involve any combination of jumping, chasing, tumbling, and "boxing". Play fighting involves rats going for each other's necks, while serious fighting involves strikes at the others' back ends. If living space becomes limited, rats may turn to aggressive behavior, which may result in the death of some animals, reducing the burden over the living space.

Rats like most mammals also form family groups, a mother and her young. This applies to both groups of males and females. However, rats are territorial animals, meaning that they usually act aggressively or scared of strange rats. Rats will fluff up their hair, hiss, squeal, and move their tails around when defending their territory.

Plate 78: Brown Rat Eating Sunflower Seeds.

Plate 79: Brown Rat in Flower Box.

Rats will chase each other, groom each other, sleep in group nests, wrestle with each other, have dominance squabbles, communicate, and play in various other ways with each other. Huddling is an additional important part of rat socialization. Huddling is often supposed to have a heat-conserving function. Nestling rats especially depend on heat from their mother, since they cannot regulate their own temperature. Huddling is an extreme form of herding. Other forms of interaction include, crawling under, which is literally the act of crawling underneath one another,walking over, also explained in the name, then there is allo-grooming, so-called to distinguish it from self-grooming. And lastly there is another type of contact called nosing, where a rat gently pushes with its nose at another rat near the neck.

Burrowing

Rats are known to burrow extensively, both in the wild and in captivity, if given access to a suitable substrate. Rats generally begin a new burrow adjacent to an object or structure, as this provides a sturdy "roof" for the section of the burrow nearest to the ground's surface. Burrows usually develop to eventually include multiple levels of tunnels, as well as a secondary entrance. Older male rats will generally not burrow, while young males and females will burrow vigorously.

Burrows provide rats with shelter and food storage, as well as safe, thermo-regulated nest sites. Rats use their burrows to escape from perceived threats in the surrounding environment; for example, rats will retreat to their burrows following a sudden, loud noise or while fleeing an intruder. Burrowing can therefore be described as a "pre-encounter defensive behavior", as opposed to a "post-encounter defensive behavior", such as flight, freezing, or avoidance of a threatening stimulus.

Distribution and Habitat

Likely originating from the plains of Asia, northern China and Mongolia, the brown rat spread to other parts of the world sometime in the middle Ages. The question of when brown rats became commensal with humans remains unsettled, but as a species, they have spread and established themselves along routes of human migration and now live almost everywhere humans are.

As it spread from Asia, the brown rat generally displaced the black rat in areas where humans lived. In addition to being larger and more aggressive, the change from wooden structures and thatched roofs to bricked and tiled buildings favored the burrowing brown rats over the arboreal black rats. In addition, brown rats eat a wider variety of foods, and are more resistant to weather extremes.

In the absence of humans, brown rats prefer damp environments, such as river banks. However, the great majority are now linked to man-made environments, such as sewage systems **(Plate 79).**

It is often said that there are as many rats in cities as people, but this varies from area to area depending on climate, living conditions, etc. Brown rats in cities tend not to wander extensively, often staying within 20 m (66 ft) of their nest if a suitable concentrated food supply is available, but they will range more widely where food availability is lower. There is great debate over the size of the population of rats in

New York City, with estimates from almost 100 million rats to as few as 250,000. Experts suggest New York is a particularly attractive place for rats because of its aging infrastructure, high moisture, and high poverty rates. In addition to sewers, rats are very comfortable living in alleyways and residential buildings, as there is usually a large and continuous food source in those areas.

In the United Kingdom, some figures show the rat population has been rising, with estimations that 81 million rats reside in the UK. Those figures would mean there are 1.3 rats per person in the country. High rat populations in the UK are often attributed to the mild climate, which allow them higher survival rates during the winter months.

The only brown rat-free zones in the world are the Arctic, the Antarctic, some especially isolated islands, the province of Alberta in Canada, and certain conservation areas in New Zealand.

Antarctica is almost completely covered by ice and has no permanent human inhabitants, making it uninhabitable by rats. The Arctic has extremely cold winters that rats cannot survive outdoors, and the human population density is extremely low, making it difficult for rats to travel from one habitation to another. When the occasional rat infestation is noticed and eliminated, the rats are unable to reinfest it from an adjacent one. Isolated islands are also able to eliminate rat populations because of low human population density and geographic distance from other rat populations.

3. House Mouse (*Mus musculus* Linnaeus)

Classification

Kingdom: Animalia

Phylum: Chordata

Class: Mammalia

Order: Rodentia

Family: Muridae

Genus: *Mus*

Species: *M. musculus*

Binomial name: *Mus musculus* **Linnaeus, 1758**

Introduction

The **house mouse** (*Mus musculus*) probably has a world distribution more extensive than any mammal apart from humans. They do considerable damage by destroying crops and consuming and/or contaminating food supplies intended for human consumption. They are prolific breeders, sometimes errupting and reaching plague proportions. They have also been implicated in the extinction of indigenous species in ecosytems they have invaded and colonised which are outside their natural range. An important factor in the success of the house mouse is their behavioural

Plate 80: House Mouse (*Mus musculus*).

plasticity brought about by the decoupling of genetics and behaviour. This enables the house mouse to adapt quickly and to survive and prosper in new environments.

The **house mouse** (***Mus musculus***) is a small mammal of the order Rodentia, characteristically having a pointed snout, small rounded ears, and a long naked or almost hairless tail. It is one of the most numerous species of the genus *Mus*. Although a wild animal, the house mouse mainly lives in association with humans.

The house mouse has been domesticated as the pet or **fancy mouse**, and as the laboratory mouse, which is one of the most important model organisms in biologyand medicine. The complete mouse reference genome was sequenced in 2002. Laboratory mice derived from the house mouse are by far the most common mammalian species used in genetically engineered models for scientific research.

Description

The house mouse is one of the most widely distributed and successful mammals in the world. It has dull greyish-brown fur and the tail, which is the same length as the body, is thicker and scalier than that of other species of mice. It is accompanied by a distinctive strong 'stale' odour and its presence can easily be detected by means of its droppings. Forms of this species living in association with man ('commensal' forms) tend to be larger and darker than 'wild' forms, and have longer tails. The voice is a familiar high-pitched 'squeak **(Plate 80).**

Biology

House mice are typically active at night, but will emerge during the day if food is scarce. They are extremely agile, with an excellent sense of balance, and are able to jump and swim fairly well. The senses of hearing and smell are highly developed and communication is largely through scent. They have an extremely broad diet, incorporating most human foodstuffs, invertebrates and occasionally more bizarre household items such as soap and tobacco. In urban situations, territories are usually set up, which males defend aggressively. Breeding tends to occur throughout the year, with five to ten litters produced each year, each one consisting of between four and eight young. The nest is constructed of shredded matter such as paper or cloth and females may share a nest if the population density is high. The young are born virtually hairless, with sealed eyes and ears. They are fully furred after 14 days and weaned after 18 - 20 days, when they begin to emerge from the nest. House mice are well-known pests, contaminating foodstuffs and grains and carrying a number of diseases and parasites that are transmissible to humans. Its close association with humans has led to it featuring widely in folklore.

Reproduction

House mice have a polygynous mating system. The recent discovery of ultrasonic songs produced by male mice, when exposed to female sex pheromones, suggests that this behavior may be involved in mate choice.

Female house mice have an **estrous cycle** about four to six days long, with estrus itself lasting less than a day. If several females are held together under crowded conditions, they will often not have an estrus at all. If they are then exposed to male urine, they will come into estrus after 72 hours.

Male house mice court females by emitting characteristic ultrasonic calls in the 30 kHz–110 kHz range. The calls are most frequent during courtship when the male is sniffing and following the female; however, the calls continue after mating has begun, at which time the calls are coincident with mounting behaviour. Males can be induced to emit these calls by female pheromones. The vocalizations appear to differ between individuals and have been compared to **bird songs** because of their complexity. While females have the capability to produce ultrasonic calls, they typically do not do so during mating behaviour.

Following copulation, female mice normally develop a **copulation plug** which prevents further copulation. The plug is not necessary for pregnancy initiation, as this will also occur without the plug. The presence or absence of the plug will not affect litter size either. This plug stays in place for some 24 hours. The **gestation** period is about 19–21 days, and they give birth to a litter of 3–14 young (average six to eight). One female can have 5 to 10 litters per year, so the mouse population can increase very quickly. Breeding occurs throughout the year. (However, animals living in the wild do not reproduce in the colder months, even though they do not hibernate.)

The pups are born blind and without fur or ears. The ears are fully developed by the fourth day, fur begins to appear at about six days and the eyes open around 13 days after birth; the pups are weaned at around 21 days. Females reach sexual maturity at about six weeks of age and males at about eight weeks, but both can copulate as early as five weeks. If the infants live in high temperatured area from birth, they will become less-haired.

Polygamy

Although house mice can be either monogamous or polygamous, they are most commonly **polygamous**. They generally show characteristics of mate-defense **polygyny** in that males are highly territorial and protective of their mates, while females are less **agonistic**. The communal nursing groups that result from these behaviors lead to lower numbers of **infanticide** since more females are able to protect greater numbers of offspring.

Mating System: polygynous

Mus musculus is characterized by tremendous reproductive potential. Breeding occurs throughout the year, although wild mice may have a reproductive season extending only from April to September. The estrous cycle is 4-6 days long, with estrus lasting less than a day. Females experience a postpartum estrus 12-18 hours after giving birth. Females generally have 5-10 litters per year if conditions are suitable, but as many as 14 have been reported. Gestation is 19-21 days but may be extended by several days if the female is lactating. Litters consist of 3-12 (generally 5 or 6) offspring, which are born naked and blind. They are fully furred after 10 days, open their eyes at 14 days, are weaned at 3 weeks, and reach sexual maturity at 5-7 weeks. Average life span is about 2 years in captivity, but individuals have lived for as long as 6 years. In the wild, most mice do not live beyond 12-18 months.

Breeding season: Throughout the year

Range number of offspring: 3.0 to 12.0.

Average number of offspring: 5.0.

Range gestation period: 21.0 (high) days.

Average weaning age: 21.0 days.

Range age at sexual or reproductive maturity (female): 5.0 to 7.0 weeks.

Range age at sexual or reproductive maturity (male): 5.0 to 7.0 weeks.

Key Reproductive Features: iteroparous; year-round breeding; gonochoric/ gonochoristic/dioecious (sexes separate); sexual; fertilization (Internal); viviparous

Average birth mass: 1.25 g.

Young mice are cared for in their mother's nest until they reach 21 days old. Soon after this most young mice leave their mother's territory, though young females are more likely to stay nearby.

Parental Investment: altricial; pre-fertilization (Protecting: Female); pre-hatching/birth (Provisioning: Female, Protecting: Female); pre-weaning/fledging (Provisioning: Female, Protecting: Female)

Breeds throughout the year under mild indoor conditions and breeds seasonally in natural habitats. Gestation lasts 19-21 days (may be longer if female is lactating). Young are weaned in about 18 days. Produces many litters of 3-12 (typically 4-5) young per year. Sexually mature in 5-10 weeks.

Range Description

Mus musculus was originally a Palaearctic species, but through its close association with humans it has been widely introduced across the globe (Musser and Carleton, 2005). The species is widespread over all continents, except Antarctica, and has become established in North and South America, sub-Saharan Africa, Australia, and many oceanic islands.

Global Distribution: Throughout the world

India: Kerala: Eravikulam NP; Periyar TR. Tamil Nadu: Mudumalai Wildlife Sanctuary, Indira Gandhi Wildlife Sanctuary, Kalakad Mundanthurai Tiger Reservem Karnataka; Nagrhole National Park, Maharashtra; Chandoli NP. House mice may originally be from Europe and Asia, from the Mediterranean region to China, but they are now distributed throughout the world by humans and live as a human commensal.

Morphology and Characteristics

House mice have an adult body length (nose to base of tail) of 7.5–10 cm (3.0–3.9 in) and a tail length of 5–10 cm (2.0–3.9 in). The weight is typically 10–25 g (0.4–0.9 oz). In the wild they vary in colour from light to dark agouti (light to dark brown), but domesticated fancy mice and laboratory mice are produced in many colors ranging from white to champagne to black. They have short hair and some, but not all, subspecies have a light belly. The ears and tail have little hair. The hind feet are short compared to**Apodemus** mice, only 15–19 mm (0.59–0.75 in) long; the normal gait is a run with a stride of about 4.5 cm (1.8 in), though they can jump vertically up to 45 cm (18 in). The voice is a high-pitched squeak. House mice thrive under a variety of conditions; they are found in and around homes and commercial structures, as well as in open fields and agricultural lands.

New born males and females can be distinguished on close examination as the anogenital distance in males is about double that of the female. From the age of about 10 days, females have five pairs of mammary glands and nipples; males have no nipples. When sexually mature, the most striking and obvious difference is the presence of testicles on the males. These are large compared to the rest of the body and can be retracted into the body.

The tail, which is used for balance, has only a thin covering of hair as it is the main peripheral organ of heat loss in **thermoregulation** along with – to a lesser extent – the hairless parts of the paws and ears. Blood flow to the tail can be precisely controlled in response to changes in ambient temperature using a system of **arteriovenous anastomoses** to increase the temperature of the skin on the tail by as much as **10 °C** to lose body heat. Tail length varies according to the environmental temperature of the mouse during postnatal development, so mice living in colder regions tend to have shorter tails. The tail is also used for balance when the mouse is climbing or running, or as a base when the animal stands on its hind legs (a behaviour known as **tripoding**), and to convey information about the dominance status of an individual in encounters with other mice.

In addition to the regular pea-sized thymus organ in the chest, house mice have a second functional pinhead-sized thymus organ in the neck next to the trachea.

Physical Description

House mice are from 65 to 95 mm long from the tip of their nose to the end of their body, their tails are 60 to 105 mm long. Their fur ranges in color from light brown to black, and they generally have white or buffy bellys. They have long tails that have very little fur and have circular rows of scales. House mice tend to have longer tails and darker fur when living closely with humans. They range from 12 to 30 g in weight. Many domestic forms of mice have been developed that vary in color from white to black and with spots.

Range mass: 12.0 to 30.0 g.

Range length: 65.0 to 95.0 mm.

Other Physical Features: endothermic; homoiothermic; bilateral symmetry

Sexual Dimorphism: sexes alike

Average basal metabolic rate: 0.271 W.

Length: 20 cm

Weight: 23 grams

Behaviour

House mice usually run, walk, or stand on all fours, but when eating, fighting, or orienting themselves, they rear up on their hind legs with additional support from the tail - a behaviour known as "tripoding". Mice are good jumpers, climbers, and swimmers, and are generally considered to be **thigmotactic,** *i.e.* usually attempts to maintain contact with vertical surfaces.

Mice are mostly **crepuscular** or **nocturnal**; they are averse to bright lights. The average sleep time of a captive house mouse is reported to be 12.5 hours per day. They live in a wide variety of hidden places near food sources, and construct nests from various soft materials. Mice are territorial, and one dominant male usually lives together with several females and young. Dominant males respect each other's territories and normally enter another's territory only if it is vacant. If two or more males are housed together in a cage, they often become aggressive unless they have been raised together from birth.

House mice primarily feed on plant matter, but are **omnivorous**. They eat their own faeces to acquire nutrients produced by bacteria in their intestines. House mice, like most other rodents, do not vomit.

Mice are generally afraid of rats which often kill and eat them, a behavior known as muricide. Despite this, free-living populations of rats and mice do exist together in forest areas in New Zealand, North America, and elsewhere. House mice are generally poor competitors and in most areas cannot survive away from human settlements in areas where other small mammals, such as wood mice, are present. However, in some areas (such as Australia), mice are able to coexist with other small rodent species.

Social Behavior

The social behaviour of the house mouse is not rigidly fixed into species-specific patterns but is instead adaptable to the environmental conditions, such as the availability of food and space. This adaptability allows house mice to inhabit diverse areas ranging from sandy dunes to apartment buildings.

House mice have two forms of social behaviour, the expression of which depends on the environmental context. House mice in buildings and other urbanized areas with close proximity to humans are known as **commensal.** Commensal mice populations often have an excessive food source resulting in high population densities and small home ranges. This causes a switch from territorial behaviour to a hierarchy of individuals. When populations have an excess of food, there is less female-female aggression, which usually occurs to gain access to food or to prevent infanticide. Male-male aggression occurs in commensal populations, mainly to defend female mates and protect a small territory. The high level of male-male aggression, with a low female-female aggression level is common in **polygamous** populations. The social unit of commensal house mouse populations generally consists of one male and two or more females, usually related. These groups breed cooperatively, with the females communally nursing. This cooperative breeding and rearing by related females helps increase reproductive success. When no related females are present, breeding groups can form from non-related females.

In open areas such as shrubs and fields, the house mouse population is known as noncommensal. These populations are often limited by water or food supply and have large territories. Female-female aggression in the noncommensal house mouse populations is much higher, reaching a level generally attributed to free-ranging species. Male aggression is also higher in non commensal populations. In commensal populations, males come into contact with other males quite frequently due to high population densities and aggression must be mediated or the risk of injury becomes too great.

Both commensal and noncommensal house mouse males aggressively defend their territory and act to exclude all intruders. Males mark their territory by scent marking with urine. In marked territories, intruders showed significantly lower aggression than the territory residents. House mice show a male-biased dispersal; males generally leave their birth sites and migrate to form new territories whereas females generally stay and are opportunistic breeders rather than seasonal.

Habitat

House mice generally live in close association with humans– in houses, barns, granaries, etc. They also occupy cultivated fields, fencerows, and even wooded areas, but they seldom stray far from buildings. Some individuals spend the summer in fields and move into barns and houses with the onset of cool autumn weather. Because of their association with humans, house mice have been able inhabit inhospitable areas (such as tundra and desert) which they would not be able to occupy independently.

Niche

Ruderal, scrub and forested tracts

Habitat Status

Quantitative and qualitative increase due to increase in human habitations and creation of suitable microhabitats

Trophic Strategy

In the wild, house mice eat many kinds of plant matter, such as seeds, fleshy roots, leaves and stems. Insects (beetle larvae, caterpillars, and cockroaches) and meat (carrion) may be taken when available. In human habitation, *Mus musculus* consumes any human food that is accessible as well as glue, soap, and other household materials. Many mice store their food or live within a human food storage facility.

Animal Foods: carrion; insects; terrestrial non-insect arthropods

Plant Foods: leaves; roots and tubers; wood, bark, or stems; seeds, grains, and nuts; fruit

Foraging Behavior: stores or caches food

Primary Diet: omnivore

Food Habits

In the wild, house mice eat many kinds of plant matter, including seeds, roots, leaves, and stems. They will also eat insects (Coleoptera, Lepidoptera, and Blattaria) and meat if it is available. If house mice live near humans, they will eat any human food that is available as well as glue, soap, and other household materials. Many mice will gather and then store their food for later use.

Known Predators

- ☆ Domestic cats (*Felis silvestris*)
- ☆ Red foxes (*Vulpes vulpes*)
- ☆ Weasels (*Mustela*)
- ☆ Ferrets (*Mustela*)
- ☆ Mongooses (*Herpestidae*)
- ☆ Large lizards (*Squamata*)
- ☆ Snakes (*Serpentes*)
- ☆ Hawks (*Accipitridae*)
- ☆ Falcons (*Falconidae*)
- ☆ Owls (*Strigiformes*)

Mice and Diseases

- ☆ House mice can sometimes transmit diseases, contaminate food, and damage food packaging.
- ☆ Lymphocytic choriomeningitis (LCMV) can be transmitted by mice, but is not a commonly reported infection in humans, though most infections are mild and are often never diagnosed
- ☆ House mice are not usually a vector of human plague (**bubonic plague**) because they have less infestations with fleas than do rats, and because the fleas which house mice normally carry exhibit little tendency to bite humans rather than their natural host.
- ☆ **Rickettsialpox**, caused by the bacterium *Rickettsia akari* and similar to **chickenpox**, is spread by mice in general, but is very rare and generally mild and resolves within 2–3 weeks if untreated. No known deaths have

resulted from the disease.**Murine typhus** (also called endemic typhus) is caused by the bacterium *Rickettsia typhi*, and is transmitted by the fleas that infest rats. While rat fleas are the most common vectors, cat fleas and mouse fleas are less common modes of transmission. Endemic typhus is highly treatable with antibiotics. The US CDC currently does not mention rickettsialpox or murine typhus on its website about diseases directly transmitted by rodents (in general).

☆ **Leptospirosis** is carried by a variety of wild and domestic animals including dogs, rats, swine, cattle, mice in general, and can be transmitted by the urine of an infected animal and is contagious as long as the urine is still moist

4. Lesser Bandicoot Rat, Sind Rice Rat or Indian Mole-Rat (*Bandicota bengalensis* Gray)

Classification

Kingdom: Animalia

Phylum: Chordata

Class: Mammalia

Order: Rodentia

Family: Muridae

Genus: *Bandicota*

Species: *B. bengalensis*

Binomial name *Bandicota bengalensis* Gray, 1835)

The **lesser bandicoot rat, Sind rice rat** or **Indian mole-rat** (*Bandicota bengalensis*) is a giant rat of Southern Asia, not related to the true bandicoots. They can be up to 40 cm long (including the tail), are considered a pest in the cereal crops and gardens of India and Sri Lanka, and emit piglike grunts when attacking. The name bandicoot is derived from the Telugu language word *pandikokku*, which translates loosely to "**pig-rat**". Like the better known rats in the genus *Rattus*, bandicoot rats are members of the family Muridae. Their furis dark or (rarely) pale brown dorsally, occasionally blackish, and light to dark grey ventrally. The head-body length is around 250 mm, and the uniformly dark tail is shorter than the head-body length.

Description

The lesser bandicoot and two other species are nocturnal or most active at twilight. They construct burrows to nest and bear their litters. The number of bandicoot babies can range from two to 18. Their staple diet is grains, fruit and invertebrates. They are prone to destroying cultivated crops in fields. Of all the three species, the lesser bandicoot is an especially aggressive burrower and has been reported to make tunnels in concrete cellars **(Plates 81-83).**

Plates 81–83: *Bandicota bengalensis* Gray, 1835.

Habitat

These rats are also known to inhabit houses in villages and are particularly aggressive when threatened. The controls are done by mechanical (mouse trap etc.), rodenticides and biological control (by introducing rodent diseases etc.)

Ecology

Commonly lives in cultivated plains and gardens and is one of the most destructive pests to crops and cultivation. It digs burrows with characteristic pile of earth around the entrance, hence its name. The burrow system is extensive and elaborate, consisting of numerous chambers (sleeping, storing etc.), galleries and exits or 'bolt-holes', which are covered with loose earth, facilitating an easy escape during emergencies. The storage chambers are stocked with large amounts of grain, especially during harvest time. Usually one mole-rat is found in one burrow, except when there is a mother with young. It has a habit of erecting its long guard hairs scattered over the back and emitting harsh grunts when disturbed.

Reproduction

Female can have up to 10 litters. Young (10-12 per litter) are born blind and naked. Young reach sexual maturity around 60 days after birth. Life span of adults is about 8-9 months.

As a Vector

It is a carrier of and spreads many diseases such as;

- ☆ Plague
- ☆ Murine typhus

- ☆ Leptospirosis
- ☆ Salmonellosis
- ☆ Rat bite fever

Susceptibility to Drugs

- ☆ **Warfarin** is a first generation anticoagulant that relies on multiple feeding events to achieve lethality in susceptible rodents. Majority of Lesser bandicoot rats are highly susceptible to warfarin, where according to one experiment, one female animal has survived a high dose of active ingredient (79.1 mg kg-1).
- ☆ **Triptolide** has been reported to cause antifertility in male rats and mice. Triptolide treatment affected the histomorphology of uterus of these rats by causing a decrease in lumen and columnar cell height and number of uterine glands and ovary by increasing the number of atretic follicles and decreasing the number of developing follicles.
- ☆ There appears to be aversion to **scilliroside** at all concentrations in food baits by lesser bandicoot rats. Maximum mortality attained on free-choice feeding on scilliroside is 90 per cent among the individuals.

5. Greater Bandicoot Rat (*Bandicota indica* Bechstein, 1800)

Classification

Kingdom: Animalia

Phylum: Chordata

Class: Mammalia

Order: Rodentia

Family: Muridae

Genus: ***Bandicota***

Species: ***B. indica***

Binomial name: ***Bandicota indica*** **Bechstein, 1800)**

The **Greater Bandicoot Rat** (***Bandicota indica***) is a species of rodent in the family Muridae. It is found in China, India, Indonesia, Laos, Malaysia, Myanmar, Nepal, SriLanka, Taiwan, Thailand, and Vietnam. It can grow to about 27–29cm without including the tail which tail can grow to 28cm.

Description

Dark gray-brown upperparts with a profusion of long black hairs. Sides gray with a few long black hairs. Short light gray fur on the ventral surface. Dark and naked scaly tail, dark feet with light-colored claws. The young are much lighter in colour **(Plates 84-85).**

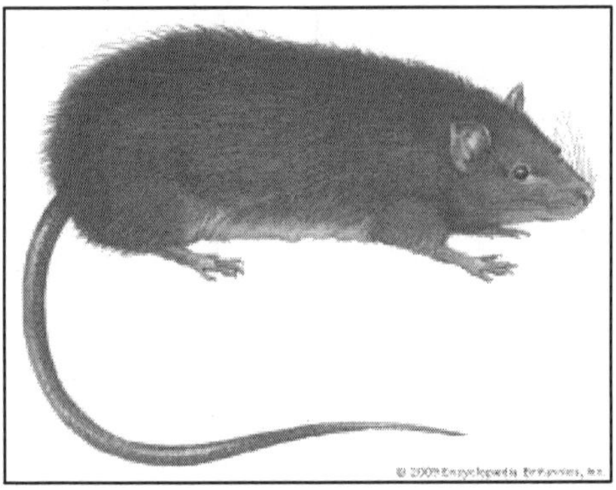

Plates 84–85: Greater Bandicoot Rat (*Bandicota indica*).

Reproduction

Female has between 8 to 10 litters. Young (8–14 per litter) are born blind and naked. Young reach sexual maturity around 50 to 60 days after birth. The life fe span of adults is around a year.

Behaviour

Large, aggressive bandicoot rats, that erect their guard hairs on their backs and emit grunts when disturbed. If caged with other bandicoots, it is likely to fight to death within a few hours. Usually occupies the outskirts of human dwellings such as compounds and gardens and is commonly found near garbage bins. Its burrowing habits cause great damage to grounds and flooring as it can also tunnel through brick and masonry. The characteristic large burrow gives away its presence. A totally non-fussy eater, it is equally comfortable feeding on household refuse, grain and vegetables and is a very serious pest in poultry farms. Like its smaller cousin, it is also a carrier for many diseases.

Parasites

Parasites of *Bandicota indica* include:

- *Schistosoma spindale*
- *Angiostrongylus cantonensis*

Graeter Bandicoot Rat Borne Disease

- Leptospirosis
- Hantavirus
- Babesiosis
- Bovine schistosomiasis.

Impact on Public Health

Greater bandicoot rat is a important vector of rat lungworm, *Angiostrongylus cantonensis* that is widespread in South-East Asia. The typical symptoms of angiostrongyliasis disease, which is caused by the third-stage larvae, resemble meningitis. Both humans and rodents can act as the primary host for the lungworm.

In Myanmar, the greater bandicoot rat was known to act as a vector for **murine typhus**.

Habitat and Ecology

This species is commonly found in fields, villages and cities. It naturally is found in swampy areas, and is especially common in lowland rice fields. Animals construct burrows in stream banks, paddy dikes and the edges of fields. The typical litter size between five and seven young.

Systems: Terrestrial

6. Indian Gerbil (*Tatera indica* Hardwicke, 1807)

Classification

Kingdom: Animalia

Phylum: Chordata

Class: Mammalia

Order: Rodentia

Family: Muridae

S. Family: Gerbillinae

Genus: *Tatera* Lataste, 1882

Species: *T. indica*

Binomial name: *Tatera indica* Hardwicke, 1807)

The **Indian gerbil, *Tatera indica,*** also known as "Antelope rat", is a species of gerbil in the family Muridae. It is found in Afghanistan,India, Iran, Iraq, Kuwait, Nepal, Pakistan, Sri Lanka, and Syria. It is the only species in the genus ***Tatera***. Members of the genus *Gerbilliscus* have historically been placed in the genus *Tatera*.

Description

Head and body length is 17-20cm. Tail is 20-21cm. Dorsal surface including entire head is light brown or light brown with rusty wash. Underparts are white. Tail fully furred, dark blackish brown with grayish sides and prominent black tuft on tip. Fur on body soft, sparse underneath; tail fur is longer. Eyes are large and prominent. Bounding gait is distinguished when running **(Plate 86).**

Distribution

Indian gerbils are prevalent all throughout south Asia, and particularly in India, Pakistan and Sri Lanka. These rodents also sometimes live in more western locations of the Middle East, such as Syria, Iran, Kuwait and Turkey. Indian gerbils can also be found in Afghanistan and Nepal.

Plate 86: Indian Gerbil.

Diet

Omnivorous. Known to eat grains, seeds, plants, roots, insects, reptiles and even small birds and mammals it can catch up. Insects make up an especially large portion of their diets in monsoon season. Indian gerbils even occasionally consume other tiny rodents.

Ecology and Habitat

This species is found in a range of dry or arid habitats. In South Asia, it occurs in dry deciduous forests, scrub forests, grasslands, rocky areas, hot deserts, arid and semi-arid regions and uncultivated areas. It has been found to occupy undisturbed barren open areas. In Turkey, it appears to prefer uncultivated arid and semi-arid habitats with soft soil and Dry River slopes. These light brown gerbils are serious about burrowing, and therefore tend to gravitate toward places that allow for it. Indian gerbils burrow for a lot of reasons, which include keeping sustenance tucked away and resting. Some typical environments for these animals are plains with ample sand, grasslands, deciduous forests, hot and arid deserts, scrub forests and very rugged regions. Indian gerbils are especially common in arid and dry climates. It is also not rare to see them in areas of farmland, particularly those that are close to water sources.

Farmers often consider Indian gerbils to be vermin or pests due to the crop destruction they cause. These gerbils often feed on grain, seeds, saplings, corn ears and sprouts when in agricultural sites.

This gerbil's preferred habitat is almost anywhere that is not too sandy, or too cold. In India burrows of Indian Gerbils have been found alongside main streets in the towns, and even in the granaries of the major cities of Pakistan. In Indian villages, they often burrow into hedgerows and mud walls that border the fields. They are very common in areas of human habitation and in Iraq and Syria it is almost only found near villages. In Iran they live more remotely in areas with green vegetation all year round. In Afghanistan they are found near to isolated buildings on the edges of semi-desert areas and in dry stream beds. As high altitude brings colder temperatures they become rarer as the ground rises. In true desert areas such as Rajasthan they avoid shifting sand dunes and prefer the firmer plains.

In the deserts of Northwest India the Indian Gerbil often lives alongside the Indian Desert Gerbil (*Meriones hurrianae*). This species prefers drier, sandy habitats. The difference between the two species can be seen by the way that *M. hurrianae* is more common in cultivated fields that are not irrigated, whilst *T. indica* is more common where the fields are irrigated.

In Culture

In Southern India and Sri Lanka, it is dug out of its burrows as a source of food, but in the north the animal is thought too small to be worth eating. The Indian Gerbil is also taboo to many Hindus as it is considered "A vehicle of Lord Ganesh"

Known Predators

Tatera indica is Prey of

Strigiformes

Serpentes

Varanus

Vulpes

Felis silvestris

Canis aureus

Falconiformes

Known prey organisms

Tatera indica Preys on

Arthropoda

Insecta

Aves

Mammalia

Life Expectancy

A captive *Tatera indica* lived for 7 years. In the wild, however, most individuals don't survive their first year of life and adults probably live only a few years.

Reproduction

Males and females of this species live apart. The relationship between the sexes is currently not known. It is not yet known whether mating occurs above or below ground in burrows. Duration of the estrous cycle was found to be 4.5 days in the laboratory. The gestation period for *T. indica* ranges from 21 to 30 days, with litter size ranging from 1 to 10 young, with 5 to 6 being the most common number of young per litter. Young Indian gerbils are independent as early as 21 days of age and reach sexual maturity as early as 10 weeks of age. Females attain sexual maturity earlier than males.

Breeding interval: Indian gerbils can breed multiple times throughout the year, the interval is not well known.

Breeding season: Indian gerbils breed throughout the year, with peaks in February, July, August, and November.

Range number of offspring: 4 to 10.

Range gestation period: 21 to 30 days.

Range time to independence: 21 (low) days.

Range age at sexual or reproductive maturity (female): 10 (low) weeks.

Range age at sexual or reproductive maturity (male): 10 (low) weeks.

Key Reproductive Features: iteroparous; year-round breeding; gonochoric/gonochoristic/dioecious (sexes separate); sexual; viviparous

Average birth mass: 3 g.

Average number of offspring: 5.4.

Parental care in *T. indica* has not been described. However, as in all mammals, females nurse and care for their young until they reach independence. The young are born in a relatively helpless state in a nest chamber in a burrow. Their eyes open at 14 days old.

Parental Investment: altricial; female parental care

7. Indian Desert Jird (Gerbil) (*Meriones hurrianae*, Jerdon, 1867)

Kingdom: Animalia

Phylum: Chordata

Class: Mammalia

Order: Rodentia

Family: Muridae

S. Family: Gerbillinae

Genus: ***Meriones***

Species: M. hurrianae

Binomial name: ***Meriones hurrianae*****(Jerdon, 1867)**

The **Indian desert jird,** ***Meriones hurrianae,*** is a species of jird found mainly in the Thar Desert in India. Jirds are closely related to gerbils.

Distribution

The Indian desert jird is found in southeastern Iran and Pakistan to northwestern India. In India they can be found in Rajasthan andGujarat. Jirds are closely related to gerbils.

Description

The Indian desert jird has a grey-brown coat with yellowish-grey belly. It is approximately 12–14 centimetres (4.7–5.5 in) long and has a tail 10–15 cm (3.9–5.9 in) long. The distinguishing characteristics include short ears, long black claws and orange incisors **(Plates 87-89).**

Habitat

Jirds inhabit desert and barren areas preferring firm soil. They are not to be found in pure sand dunes or rocky outcrops.

Habits

The jirds are gregarious and their burrows are seen close by. Each jird will have at least two or more entrances to his burrow complex. Often the entrance is in the

Plates 87–89: Desert Jird *Meriones hurrianae.*

shade of a tree or near the trunk of bushes. They feed on seeds, roots, nuts, grasses and insects. This is a diurnal, fossorial, social and colonial (5-20 individuals) species. The species forms superficial to deep burrows in plains, prefers hammocky landscapes on sandy plains with higher density of bushes. Typical habitats include alluvial plains, uncultivated clay flats, sandy plains, interdunal regions, gravelly depressions with grasses and other vegetation.The species can also be found in acacia forest and hedges. It has been found to occupy the edges of cultivated fields.

Natural History

They construct burrows to aid in food storage, temperature regulation and water retention. The burrows of some species are rather simple, but others can be quite complex. *Meriones crassus* adults have been reported to have burrows with a combined length of over 30 meters and have 18 entrances.[2] Food is stored in chambers of the burrows. As with other arid adapted rodents, stored food has the capacity to reabsorb moisture given off by the animal during respiration. Jirds can go their entire lives without drinking, relying instead on water generated during metabolism. These rodents feed on roots, seeds, fruits, and insects.

Members of the genus range from being social to rather solitary. Even in solitary species, home ranges often overlap. Females give birth to one to 12 young after a gestation period of about 20–30 days.**Sexual maturity** is reached after about 9–15 weeks. **Longevity** in the wild is usually less than six months, but the record for a captive animal is over five years.

Difference between Jirds and gerbils

Jirds are closely related to gerbils but are differentiated by the absence of long hind feet and the characteristic erect posture of a gerbil. The tail is generally shorter than the head and body in a jird. It also has much shorter ears. In addition, the Indian desert jird is diurnal, in comparison to the three common gerbil species found in India which are nocturnal.

Threats

This species is threatened in parts of its range by habitat loss and degradation due to expansion of agriculture, small-scale logging operations, invasions of alien plant species directly impacting the habitat, collection of fuel wood. It is also threatened by pest control programmes, natural disasters (such as drought and floods) and harvest for local consumption.

8. Little Indian Field Mouse (*Mus booduga* Gray, 1837)

Classification

Kingdom: Animalia

Phylum: Chordata

Class: Mammalia

Order: Rodentia

Family: Muridae

Genus: *Mus*

Sub Genus: *Mus*

Species: M. booduga

Binomial name: ***Mus booduga*** **(Gray, 1837)**

The **little Indian field mouse** (***Mus booduga***) is a species of rodent in the family Muridae. It is found in India, Myanmar, Nepal, and Sri Lanka. A single member of the species has also been sighted in the South End of Boston, MA.

Description

Head and body length is 7cm. Tail is 6cm. Upperparts a glossy light brown fading to grayish white or white on the ventral surface. There is often a light brown band or splotch across the chest. Large rounded ears set on the head. Muzzle rather pointed. Tail is dark above, and paler below. Upper incisors curve backwards **(Plate 90).**

Plate 90: Little Indian Field Mouse (*Mus booduga*).

Population

It is reported to be a common and abundant species in India. It is also abundant in Myanmar

Habitat and Ecology

This species is common in ricefields and other irrigated croplands in India. It occurs in tropical and subtropical dry deciduous forests (Molur *et al.*, 2005). In Myanmar, it lives in drier, sandy habitat and scrubby areas, and is a serious pest in cropland areas (such as peanut crops).

Threats. There are no major threats to the species.

8. Indian Crested Porcupine/Indian Porcupine (*Hystrix indica* Kerr, 1792)

Classification

Kingdom: Animalia

Phylum: Chordata

Class: Mammalia

Order: Rodentia

Family: Hystricidae

Genus: ***Hystrix***

Sub Genus: ***Hystrix***

Species: M. booduga

Binomial name: ***Hystrix indica*****Kerr, 1792**

The **Indian crested porcupine** (*Hystrix indica*), or **Indian porcupine**, is a large species of Hystricomorph rodent (order Rodentia) belonging to the Old World porcupine family, Hystricidae. It is native to southern Asia and the Middle East.

Description

The Indian crested porcupine is a large rodent, weighing 11–18 kg. Their body (from the nose to the base of the tail) measures between 70–90 cm, with the tail adding an additional 8–10 cm. The lifespan of wild Indian crested porcupines is unknown, but the oldest known captive individual was a female that lived to be 27.1 years old **(Plates 91-92).**

It is covered in multiple layers of modified hair called quills, with longer, thinner quills covering a layer of shorter, thicker ones. The quills are brown or black with alternating white and black bands. They are made of keratin and are relatively flexible. Each quill is connected to a muscle at its base, allowing the porcupine to raise its quills when it feels threatened. The longest quills are located on the neck and shoulder, where the quills form a "skirt" around the animal. These quills can grow up to 51 cm (20 in) long, with most measuring between 15–30 cm. Smaller (20 cm) and more rigid quills are packed densely on the back and rump. These smaller quills are used to stab at potential threats. The base of the tail contains shorter quills

Plate 91: ***Hystrix indica*** **Eating Plant Material.**

Plate 92: ***Hystrix indica*** **in a Trap.**

that appear white in color, with longer, hollow quills that the porcupine can rattle to produce a warning sound when threatened. Contrary to popular belief, Indian crested porcupines (like all porcupines) cannot shoot their quills.

The Indian crested porcupine has a stocky build with a low surface area to volume ratio, which aids in heat conservation. It has broad hands and feet with long claws used for burrowing. Like all porcupines, the Indian crested porcupine has a good sense of smell and sharp, chisel-like incisors.

Habitat and Distribution

Indian crested porcupines are found throughout southwest and central Asia and in parts of the Middle East, including Afghanistan, Armenia, Azerbaijan, China, Georgia, India, Iran, Iraq, Israel, Jordan, Kazakhstan, Nepal, Pakistan, Saudi Arabia, Sri Lanka, Turkey, Turkmenistan, and Yemen. Due to their flexible environmental tolerances, Indian crested porcupines occupy a broad range of habitats. They prefer rocky hillsides, but are also common in tropical and temperate shrub-lands, grasslands, forests, plantations, and gardens. Their range seems to be limited by seasonal densities of forage and the availability of suitable substrates for digging burrows. More specifically, the northern range of the Indian crested porcupine is limited by minimum summer night duration: they do not occur above latitudes where minimum night duration is less than 7 hours, presumably because of the amount of foraging time required to meet their dietary needs.

Diet

Indian crested porcupines have a very broad and mostly herbivorous diet. They consume a variety of natural and agricultural plant material, including roots, bulbs, fruits, grains, and tubers, along with insects and small vertebrates. Because they are cecal digesters, they are able to exploit low quality forage. They have also been known to chew on bones to acquire minerals, such as calcium, that aid in quill growth. Their capability to form substantial fat reserves is a useful adaptation for living in seasonally fluctuating habitats.

These porcupines can act as substantial habitat modifiers when excavating for tubers. They are also considered serious agricultural pests in many parts of their range due to their taste for agricultural crops. For these reasons, they are often regarded as a nuisance.

Behaviour

Like other Old World porcupines, the Indian crested porcupine is nocturnal. Both adults and weaned juveniles spend an average of 7 hours foraging every night. They tend to avoid moonlight in the winter months, which could be a strategy to evade predation. However, during summer months they do not avoid moonlight (likely because there are less dark hours during which to forage) but instead tend to stay closer to their dens. During the day, they remain in their dens, but throughout the winter they will occasionally emerge from their dens during daylight hours to bask in the sun.

The Indian crested porcupine is semi-fossorial. They live in natural caves or in excavated burrows. Because they do not climb or jump well, they spend most of their life on or under the ground. However, **they are good swimmers**.

Predators of the Indian crested porcupine include large cats, caracals, wolves, striped hyenas, and humans. When excited or scared, a porcupine will stand its quills up to appear larger. It can also rattle the hollow quills at the base of its tail, stomp its feet, growl, grunt, or charge backward into the threat.

Reproduction

Due to its adaptability to a wide range of habitats and food types, the Indian crested porcupine is listed by the IUCN asLeast Concern as of 2008. Populations are stable and not severely fragmented, and while population status varies across its range, in many places it is common enough to be considered a pest. However, as a result of urbanization, infrastructure development, and pesticide use, suitable porcupine habitat is currently declining.

Management

The Indian crested porcupine is protected under the India Schedule IV of the Indian Wildlife Protection Act of 1972, amended up to 2002. Nonetheless, because they are destructive to gardens and agricultural crops, porcupines are widely hunted. There exists a large trade of these porcupines for consumption and medicinal use. Despite being considered a pest, Indian crested porcupines play an important role in spreading seeds and pollen.

10. Indian Palm Squirrel or Three-Striped Palm Squirrel (*Funambulus palmarum* Linnaeus, 1766)

Classification

Kingdom: Animalia

Phylum: Chordata

Class: Mammalia

Order: Rodentia

Family: Sciuridae

Genus: *Funambulus*

Sub Genus: *Funambulus*

Species: F. palmarum

Binomial name: *Funambulus palmarum*(Linnaeus, 1766)

Introduction

The **Indian palm squirrel** or **three-striped palm squirrel** (*Funambulus palmarum*) is a species of rodent in the family Sciuridae found naturally inIndia (south of the Vindhyas) and Sri Lanka. In the late 19th century, the palm squirrel was introduced to Madagascar, Réunion, Mayotte, Comoro Islands, Mauritius, Seychelles and Australia, where it has since become a minor pest. The closely related five-striped palm squirrel, ***F. pennantii,*** is found in northern India, and its range partly overlaps with this species.

Description

The palm squirrel is about the size of a large chipmunk, with a bushy tail slightly shorter than its body. The back is a grizzled, grey-brown colour with three conspicuous white stripes which run from head to tail. The two outer stripes run from the forelegs to the hind legs only. It has a creamy-white belly and a tail covered with interspersed, long, black and white hair. The ears are small and triangular. Juvenile squirrels have significantly lighter coloration, which gets progressively darker as they age. **Albinism** is rare, but exists in this species **(Plate 93-95).**

Life Cycle

The **gestation** period is 34 days; breeding takes place in grass nests during the autumn. Litters of two or three are common and average 2.75. The young are weaned

Plate 93: An Indian Palm Squirrel.

Plate 94: An Indian Palm Squirrel with its Pup.

Plate 95: An Indian Palm Squirrel Nest.

after about 10 weeks and are sexually mature at 9 months. Adult weight is 100 g. Little is known about their longevity, but one specimen lived 5.5 years in captivity.

Diet and Behavior

These squirrels eat mainly nuts and fruits. They are fairly vocal, with a cry that sounds like "**chip chip chip**" when danger is present. They are opportunists in urban areas, and can be easily domesticated and trained to accept food from humans. Naturally active, their activity reaches levels of frenzy during the mating season. They tend to be very protective of their food sources, often guarding and defending them from birds and other squirrels. Unlike some other species of squirrel, the Indian palm squirrel does not hibernate.

Importance in Hinduism

Squirrels are considered sacred in India and are not to be harmed. They are even fed by many Hindu families, mainly because of their association with Lord Rama.

A legend explains the stripes on the back of most of the squirrels. During the construction of the Rama Setu (bridge) atRameswaram by Lord Rama and theVanara Sena, a little squirrel also contributed in its own little way. It rolled in the beach sand and then ran to the end of the bridge to shake off the sand from its back (chanting Lord Rama's name all along).

Lord Rama, pleased by the creature's dedication, caressed the squirrel's back and ever since, the Indian squirrel carried white stripes on its back, which are believed to be the mark of Lord Rama's fingers. Lord Rama and the squirrel are mentioned in one of the hymns of the Alvars.

Management

Elimination of Breeding Grounds

- ✰ Proper waste management and drainage of still water, eliminates the breeding ground of many pests.
- ✰ Garbage provides food and shelter for many unwanted organisms, as well as an area where still water might collect and be used as a breeding ground by mosquitoes. Communities that have proper garbage collection and disposal, have far less of a problem with rats, cockroaches, mosquitoes, flies and other pests than those that don't.
- ✰ Open air sewers are ample breeding ground for various pests as well. By building and maintaining a proper sewer system, this problem is eliminated.
- ✰ Certain spectrums of LED light can "disrupt insects' breeding".

Field Burning

Traditionally, after a sugar cane harvest, the fields are all burned, to kill off any rodents, insects or eggs that might be in the fields.

Hunting

In some nations, teams of **rat-catchers** work at chasing rats from the field, and killing them with dogs and simple hand tools.

Traps

A variety of **mouse traps** and **rat traps** are available for mice and rats, including **snap traps, glue traps** and **live catch traps.**

Pesticides

Under iInsecticide act, 1968 only seven insecticides are registered as rodenticides. Among them only four rodenticides are available in the market. These are

1. Zinc phosphide
2. Aluminiul phosphide
3. Coumatetralyl and
4. Bromadiolane.

1. Metal Phosphides

Metal phosphides have been used as a means of killing rodents and are considered single-dose fast acting rodenticides (death occurs commonly within 1–3 days after single bait ingestion). Bait consisting of food and a phosphide (usually zinc phosphide) is left where the rodents can eat it. The acid in the digestive system of the rodent reacts with the phosphide to generate the toxic phosphine gas. This method of vermin control has possible use in places where rodents are resistant to some of the anticoagulants, particularly for control of house and field mice; zinc phosphide baits are also cheaper than most second-generation anticoagulants, so that sometimes, in the case of large infestation by rodents, their population is initially reduced by copious amounts of zinc phosphide bait applied, and the rest of population that survived the initial fast-acting poison is then eradicated by prolonged feeding on anticoagulant bait. Inversely, the individual rodents, that survived anticoagulant bait poisoning (rest population) can be eradicated by pre-baiting them with nontoxic bait for a week or two (this is important to overcome bait shyness, and to get rodents used to feeding in specific areas by specific food, especially in eradicating rats) and subsequently applying poisoned bait of the same sort as used for pre-baiting until all consumption of the bait ceases (usually within 2–4 days). These methods of alternating rodenticides with different modes of action gives actual or almost 100 per cent eradications of the rodent population in the area, if the acceptance/palatability of baits are good (*i.e.*, rodents feed on it readily).

2. Zinc Phosphide

Zinc phosphide is typically added to rodent baits in a concentration of 0.75 per cent to 2.0 per cent. The baits have strong, pungent garlic-like odor due to the phosphine liberated by **hydrolysis**. The odor attracts (or, at least, does not repel) rodents, but has an repulsive effect on other mammals. Birds, notably wild turkeys, are not sensitive to the smell, and will feed on the bait, and thus become collateral damage.

The tablets or pellets (usually **aluminium, calcium or magnesium phosphide** for fumigation/gassing) may also contain other chemicals which evolve ammonia, which helps to reduce the potential for spontaneous **combustion** or **explosion** of the phosphine gas.

Metal phosphides do not accumulate in the tissues of poisoned animals, so the risk of **secondary poisoning** is low.

Before the advent of **anticoagulants**, phosphides were the favored kind of rat poison. During World War II, they came into use in United States because of shortage of **strychnine** due to the Japanese occupation of the territories where the **strychnine tree** is grown. Phosphides are rather fast-acting rat poisons, resulting in the rats dying usually in open areas, instead of in the affected buildings.

Phosphides used as rodenticides include:

- Aluminium phosphide (fumigant only)
- Calcium phosphide (fumigant only)
- Magnesium phosphide (fumigant only)
- Zinc phosphide (bait only)

Hypercalcemia

Calciferols (vitamins D), **cholecalciferol** (vitamin D_3) and **ergocalciferol** (vitamin D_2) are used as rodenticides. They are toxic to rodents for the same reason they are important to humans: they affect calcium and phosphate **homeostasis** in the body. Vitamins D are essential in minute quantities (few IUs per kilogram body weight daily, only a fraction of a milligram), and like most fat **soluble vitamins**, they are toxic in larger doses, causing **hypervitaminosis**. If the poisoning is severe enough (that is, if the dose of the toxin is high enough), it leads to death. In rodents that consume the rodenticidal bait, it causes **hypercalcemia**, raising the calcium level, mainly by increasing calcium absorption from food, mobilising bone-matrix-fixed calcium into ionised form (mainly monohydrogencarbonate calcium cation, partially bound to plasma proteins, $[CaHCO_3]^+$), which circulates dissolved in the **blood plasma**. After ingestion of a lethal dose, the free calcium levels are raised sufficiently that **blood vessels, kidneys**, the **stomach** wall and **lungs** are mineralised/calcificated (formation of calcificates, crystals of calcium salts/complexes in the tissues, damaging them), leading further to heart problems (myocardial tissue is sensitive to variations of free calcium levels, affecting both myocardial contractibility and excitation propagation between atrias and ventriculas), bleeding (due to capillary damage) and possibly kidney failure. It is considered to be single-dose, cumulative (depending on concentration used; the common 0.075 per cent bait concentration is lethal to most rodents after a single intake of larger portions of the bait) or sub-chronic (death occurring usually within days to one week after ingestion of the bait). Applied concentrations are 0.075 per cent cholecalciferol and 0.1 per cent ergocalciferol when used alone, wihich can kill a rodent or a rat.

There is an important feature of calciferols toxicology, that:

- **They** are **synergistic** with **anticoagulant** toxicants, that means, that mixtures of anticoagulants and calciferols in same bait are more toxic than a sum of toxicities of the anticoagulant and the calciferol in the bait, so that a massive hypercalcemic effect can be achieved by a substantially lower calciferol content in the bait, and vice versa, a more pronounced anticoagulant/

hemorrhagic effects are observed if the calciferol is present. This synergism is mostly used in calciferol low concentration baits, because effective concentrations of calciferols are more expensive than effective concentrations of most anticoagulants.

- ☆ The first application of a calciferol in rodenticidal bait was in the Sorex product Sorexa D (with a different formula than today's Sorexa D), back in the early 1970s, which contained 0.025 per cent warfarin and 0.1 per cent ergocalciferol. Today, Sorexa CD contains a 0.0025 per cent difenacoum and 0.075 per cent cholecalciferol combination. Numerous other brand products containing either 0.075-0.1 per cent calciferols (*e.g.* Quintox) alone or alongside an anticoagulant are marketed.

3. Coumatetralyl

Coumatetralyl is an **anticoagulant** of the 4-hydroxycoumarin vitamin K antagonist type used as a rodenticide. Coumatetralyl is commonly used with grains and other cereals as a rodent poison in conjunction with a **tracking powder** to monitor feeding activity in a particular area. Tracking powder also clings to fur, which allows more poison to be ingested from grooming. Concentrations of the chemical are usually 500 mg per 1 kg of bait.

Toxicity

Symptoms of over exposure relate to failure of the **blood clotting** mechanism and include bleeding gums and failure of blood clotting after skin wounds. After one exposure the toxicity of coumatetralyl is relatively low; however, if overexposure continues for several days the product becomes more toxic. The product must therefore be constantly present in the bloodstream for more than one to two days in order to be highly toxic. A single exposure, even though relatively large, may not produce toxic symptoms as the compound is quite rapidly metabolized.Chronic animal studies show no evidence of carcinogenic or **teratogenic** effects.

Treatment: Vitamin K_1 (**phylloquinone**) is antidotal.

4. Bromadiolane

Bromadiolone is a potent **anticoagulant rodenticide**. It is a second-generation 4-hydroxy coumarin derivative and vitamin K antagonist, often called a "**super-warfarin**" for its added potency and tendency to accumulate in the liver of the poisoned organism. When first introduced to the UK market in 1980, it was effective against the populations that had become resistant to the first generation anticoagulants.

The product may be used both indoors and outdoors for rats and mice.

Toxicity

Bromadiolone can be absorbed through the digestive tract, through the lungs, or through skin contact. The pesticide is generally given orally. The substance is a vitamin K antagonist. The lack of vitamin K in the circulatory system reduces blood clotting and will cause death due to internal hemorrhaging.

Poisoning doesn't show up for 24 to 36 hours after poison is eaten and often it may take 2–5 days for the signs to show up.

Following are acute LD_{50} values for various animals (mammals):

- ✰ Rats 1.125 mg/kg b.w.
- ✰ Mice 1.75 mg/kg b.w.
- ✰ Rabbits 1 mg/kg b.w.
- ✰ Dogs > 10 mg/kg b.w. (oral MTD)
- ✰ Cats > 25 mg/kg b.w. (oral MTD)

Antidote: Vitamin K1 is used as antidote.

Pest control: Rodenticides

Anticoagulants/ Vitamin K antagonists	**Coumarins/4-Hydroxycoumarins**	**1st generation**	Warfarin, Coumatetralyl
		2nd generation	Brodifacoum, **Bromadiolone**, Difenacoum, Flocoumafen
	1,3-Indandiones	Chlorophacinone, Pindone, Diphacinone	
	Others	Difethialone is considered a second generation anticoagulant rodenticide	
Convulsants	Crimidine, Phenylsilatrane, Strychnine, Tetramethylenedisulfotetramine		
Calciferols	Cholecalciferol, Ergocalciferol		
Inorganic compounds	Aluminium phosphide, Arsenic, Barium carbonate, Calcium phosphide, Cyanide, Thallium sulfate, Zinc phosphide		
Organochlorine	Chloralose, Endrin		
Organophosphorus	Phosacetim		
Metabolic poisons	Bromethalin, Fluoroacetamide, 1,3-Difluoro-2-propanol (Gliftor), Sodium fluoroacetate		
Others	α-Naphthylthiourea, Norbormide, Pyrinuron, Scilliroside		
Indirect	Sometimes, anticoagulant rodenticides are potentiated by an **antibiotic** or **bacteriostatic agent**, most commonly **sulfaquinoxaline**. The aim of this association is that the antibiotic suppresses intestinal **symbiotic microflora**, which are a source of vitamin K. Diminished production of vitamin K by the intestinal microflora contributes to the action of anticoagulants. Added **vitamin D** also has a synergistic effect with anticoagulants.		

1. Poisoned Bait

Poisoned bait is a common method for controlling rat populations, however is not as effective when there are other food sources around, such as garbage. Poisoned meats have been used for centuries for killing off wolves, birds that were seen to threaten crops, and against other creatures. This can be a problem, since a carcass which has been poisoned will kill not only the targeted animal, but also every other animal which feeds on the carcass. Humans have also been killed by coming in contact with poisoned meat, or by eating an animal which had fed on a poisoned carcass. This tool is also used to manage several caterpillars *e.g.* Spodoptera litura, fruit flies, snails and slugs, crabs etc.

2. Anticoagulants

Anticoagulants are defined as chronic (death occurs one to two weeks after ingestion of the lethal dose, rarely sooner), single-dose (second generation) or multiple-dose (first generation) rodenticides, acting by effective blocking of the vitamin K cycle, resulting in inability to produce essential blood-clotting factors – mainly coagulation factors II (**prothrombin**) and VII **(proconvertin).**

In addition to this specific metabolic disruption, massive toxic doses of 4-hydroxycoumarin, 4-thiochromenone and **indandione** anticoagulants cause damage to tiny blood vessels (capillaries), increasing their permeability, causing diffuse internal bleeding. These effects are gradual, developing over several days. In the final phase of the intoxication, the exhausted rodent collapses due to **hemorrhagic shock** or severe **anemia** and dies calmly. The question of whether the use of these rodenticides can be considered humane has been raised.

The main benefit of anticoagulants over other poisons is that the time taken for the poison to induce death means that the rats do not associate the damage with their feeding habits.

- ☆ First generation rodenticidal anticoagulants generally have shorter elimination half-lives, require higher concentrations (usually between 0.005 per cent and 0.1 per cent) and consecutive intake over days in order to accumulate the lethal dose, and are less toxic than second generation agents.
- ☆ Second generation agents are far more toxic than first generation. They are generally applied in lower concentrations in baits – usually on the order of 0.001 per cent to 0.005 per cent – are lethal after a single ingestion of bait and are also effective against strains of rodents that became resistant to first generation anticoagulants; thus, the second generation anticoagulants are sometimes referred to as "**superwarfarins**".

Vitamin K_1 has been suggested, and successfully used, as antidote for pets or humans accidentally or intentionally exposed to anticoagulant poisons. Some of these poisons act by inhibiting liver functions and in advanced stages of poisoning, several **blood-clotting** factors are absent, and the volume of circulating blood is diminished, so that a blood transfusion (optionally with the clotting factors present) can save a person who has been poisoned, an advantage over some older poisons.

Sterilization

In 2013, New York City tested sterilization traps in a $1.1 million study. The result was a 43 per cent reduction in rat populations. The Chicago Transit Authority plans to test sterilization control in spring 2015. The sterilization method doesn't poison the rats or humans.

Natural Rodent Control

Several wildlife rehabilitation organizations encourage natural form of rodent control through exclusion and predator support and preventing secondary poisoning altogether.

Repellents

- ☆ Balsam fir oil from the **tree *Abies balsamea*** is an EPA approved non-toxic rodent repellent.
- ☆ ***Acacia polyacantha* subsp. *campylacantha*** root emits chemical compounds that repel animals including **crocodiles,snakes** and **rats**.

Chapter 6

Birds: Grainivorous and Frugivorous Bird Pests

A. Grainivorous Bird Pests

1. Common Myna (*Acridotheres tristis* Linnaeus, 1766)

Clasification

Kingdom: Animalia

Phylum: Chordata

Class: Aves

Order: Passeriformes

Family: Sturnidae

Genus: *Acridotheres*

Species: *A. tristis*

Binomial name: *Acridotheres tristis* (Linnaeus, 1766)

The **common myna** (*Acridotheres tristis*), sometimes spelled **mynah**, also sometimes known as "Indian myna", is a member of the family Sturnidae (starlings and mynas) native to Asia. An omnivorous open woodland bird with a strong territorial instinct, the myna has adapted extremely well to urban environments.

The range of the common myna is increasing at such a rapid rate that in 2000 the **IUCN Species Survival Commission** declared it one of the world's most **invasive species** and one of only three birds in the top 100 species that pose an impact to biodiversity, agriculture and human interests.

Plate 96: *Acridotheres tristis* Adult.

Plate 97: Immature at Nest.

Description

The common myna is readily identified by the brown body, black hooded head and the bare yellow patch behind the eye. The bill and legs are bright yellow. There is a white patch on the outer primaries and the wing lining on the underside is white. The sexes are similar and birds are usually seen in pairs.

The common myna obeys **Gloger's rule** in that the birds from northwest India tend to be paler than their darker counterparts in South India. **(Plate 96-97)**

Morphometry

☆ Body length: 23 centimetres (9.1 in)

Parameter/Sex	*Male*	*Female*
Average weight (gms)	109.8	120-138
Wing chord (mm)	138-153	138-147
Bill (mm)	25-30	25-28
Tarsus (mm)	34-42	35-41
Tail (mm)	81-95	79-96

Distribution

It is a species of bird native to **Asia** with its initial home range spanning from **Iran, Pakistan, India, Nepal, Bhutan,Bangladesh** and **Sri Lanka**; as well as Afghanistan, Uzbekistan,Tajikistan, Turkmenistan, Myanmar, to Malaysia,Singapore, peninsular Thailand, Indo-China and China. The myna has been introduced in many other parts of the world such as Canada, Australia, Israel, New Zealand, New Caledonia, Hawaii, South Africa, Kazakhstan, Kyrgyzstan, Florida,Uzbekistan and islands in the Indian Ocean (Seychelles, Mauritius, Réunion, Madagascar, Maldives, Andaman and Nicobar Islands and Lakshadweep archipelago) and also in islands of the Atlantic, such as Ascension and St Helena,and Pacific Oceans.

Vocalisation

The **calls** includes **croaks, squawks, chirps, clicks, whistles and 'growls'**, and the bird often fluffs its feathers and bobs its head in singing. The common myna screeches warnings to its mate or other birds in cases of predators in proximity or when it is about to take off flying. Common mynas are popular as cage birds for their singing and "speaking" abilities. Before sleeping in communal roosts, mynas vocalise in unison, which is known as "communal noise".

Breeding

Common mynas are believed to pair for life. They breed through much of the year depending on the location, building their nest in a hole in a tree or wall. They breed from sea-level to 3000 m in the **Himalayas**.

The normal clutch size is 4–6 eggs. The average size of the egg is 30.8 x 21.99 mm. The incubation period is 17 to 18 days and fledging period is 22 to 24 days. The **Asian koel** is sometimes **brood parasitic** on this species. Nesting material used by mynas include twigs, roots, tow and rubbish. Mynas have been known to use tissue paper, tin foil and sloughed off snake-skin.

During the breeding season, the daytime activity-time budget of common myna in Pune in April to June 1978 has been recorded to comprise the following: nesting activity (42 per cent), scanning the environment (28 per cent), locomotion (12 per cent), feeding (4 per cent), vocalisation (7 per cent) and preening-related activities, interactions and other activities (7 per cent).

The common myna uses the nests of woodpeckers, parakeets, etc. and easily takes to nest boxes; it has been recorded evicting the chicks of previously nesting pairs by holding them in the beak and later sometimes not even using the emptied nest boxes. This aggressive behaviour contributes to its success as an invasive species.

Food and Feeding

Like most starlings, the common myna is omnivorous. It feeds on **insects, arachnids, crustaceans, reptiles, small mammals, seeds, grain and fruits** and discarded waste from human habitation. It forages on the ground among grass for insects, and especially for **grasshoppers**, from which it gets the generic name *Acridotheres*, **"grasshopper hunter".** It however feeds on a wide range of insects, mostly picked from the ground. It is a cross-pollinator of flowers such as **Salmalia** and **Erythrina**. It walks on the ground with occasional hops and is an opportunistic feeder on the insects disturbed by grazing cattle as well as fired grass fields.

Roosting Behaviour

Common mynas roost communally throughout the year, either in pure or mixed flocks with jungle mynas, rosy starlings, house crows, jungle crows, cattle egretsand rose-ringed parakeets and other birds. The roost population can range from less than one hundred to thousands. The time of arrival of mynas at the roost starts before and ends just after sunset. The mynas depart before sunrise. The time and timespan of arrival and departure, time taken for final settlement at the roost, duration of communal sleep, flock size and population vary seasonally.

The function of **communal roosting** is to synchronise various social activities, avoid predators, exchange information about food sources.

Communal displays (pre-roosting and post-roosting) consist of aerial maneuvers which are exhibited in the pre-breeding season (November to March). It is assumed that this behaviour is related to pair formation.

Habitat

This abundant passerine is typically found in open woodland, cultivation and around habitation. Although this is an adaptable species, its population has been decreasing significantly in Singapore and Malaysia (where it is locally called as*gembala kerbau*, literally 'buffalo shepherd') due to competition with its cousin, the introduced **Javan myna**.

Invasive Species

The common myna is a pest in South Africa, North America, the Middle East, Australia, New Zealand and many Pacific islands. It is particularly problematic in Australia. Several methods have been tried to control the bird's numbers and protect native species.

Effect on Ecosystems and Humans

Threat to Native Birds

The common myna is a hollow-nesting species; that is, it nests and breeds in protected hollows found either naturally in trees or artificially on buildings (for example, recessed windowsills or low eaves). Compared to native hollow-nesting species, the common myna is extremely aggressive, and breeding males will actively defend areas ranging up to 0.83 hectares in size (though males in densely populated urban settings tend to only defend the area immediately surrounding their nests).

This aggressiveness has enabled the common myna to displace many breeding pairs of native hollow-nesters, thereby reducing their reproductive success. In Australia, their aggressiveness has enabled them to chase native birds as large as **galahs** out of their nests.

The common myna is also known to maintain up to two roosts simultaneously; a temporary summer roost close to a breeding site (where the entire local male community sleeps during the summer, the period of highest aggression), and a permanent all-year roost where the female broods and incubates overnight. Both male and female common mynas will fiercely protect both roosts at all times, leading to further exclusion of native birds.

Threat to Crops and Pasture

The common myna (which feeds mostly on ground-dwelling insects, tropical fruits such as grapes, plums and some berries and, in urban areas, discarded human food) poses a serious threat to Australian blueberry crops, though its main threat is to native bird species.

In Hawaii, where the common myna was introduced to control pest **armyworms** and **cutworms** in **sugarcane** crops, the bird has helped to spread the robust ***Lantana*** *camara* weed across the islands' open grasslands. It also has been recorded as the fourth-ranking avian pest in the fruit industry by a 2004 survey of the Hawaiian Farm Bureau and the sixth in number of complaints of avian pests overall.

2. House Sparrow (*Passer domesticus* Linnaeus, 1758)

Classification

Kingdom: Animalia

Phylum: Chordata

Class: Aves

Order: Passeriformes

Family: Passeridae

Genus: *Passer*

Species: *P. domesticus*

Binomial name: *Passer domesticus* (Linnaeus, 1758)

The **house sparrow** (*Passer domesticus*) is a bird of the sparrow family Passeridae, found in most parts of the world. A small bird, it has a typical length of 16 centimetres (6.3 inches) and a mass of 24–39.5 grams (0.85–1.39 ounces). Females and young birds are coloured pale brown and grey, and males have brighter black, white, and brown markings. One of about 25 species in the genus*Passer*, the house sparrow is native to most of Europe, the Mediterranean region, and much of Asia. Its intentional or accidental introductions to many regions, including parts of Australia, Africa, and the Americas, make it the most widely distributed wild bird.

The house sparrow is strongly associated with human habitations, and can live in urban or rural settings. Though found in widely varied habitats and climates, it typically avoids extensive woodlands, grasslands, and deserts away from human development. It feeds mostly on the seeds of grains and weeds, but it is an opportunistic eater and commonly eats insects and many other foods. Its predators include **domestic cats, hawks, owls**, and many other predatory birds and mammals **(Plates 98-100).**

Plate 98: Male House Sparrows in Breeding.

Plate 99: Male House Sparrows in Non-breeding.

Plate 100: Male House Sparrow's Head Plumage.

Because of its numbers, ubiquity and association with human settlements, the house sparrow is culturally prominent. It is extensively, and usually unsuccessfully, persecuted as an agricultural pest, but it has also often been kept as a pet as well as being a food item and a symbol of lust and sexual potency, as well as of commonness and vulgarity. Though it is widespread and abundant, its numbers have declined in some areas.

Measurements and Shape

The house sparrow is typically about 16 cm (6.3 in) long, ranging from 14 to 18 cm (5.5 to 7.1 in).[3] It is a compact bird with a full chest and a large rounded head. Its bill is stout and conical with a culmen length of 1.1–1.5 cm (0.43–0.59 in), strongly built as an adaptation for eating seeds. Its tail is short, at 5.2–6.5 cm (2.0–2.6 in) long. The wing chord is 6.7–8.9 cm (2.6–3.5 in), and the tarsus is 1.6–2.5 cm (0.63–0.98 in). In mass, the house sparrow ranges from 24 to 39.5 g (0.85 to 1.39 oz). Females usually are slightly smaller than males. The median mass on the European continent for both sexes is about 30 g (1.1 oz), and in more southerly subspecies is around 26 g (0.92 oz). Younger birds are smaller, males are larger during the winter, and females are larger during the breeding season. Birds at higher latitudes, colder climates, and sometimes higher altitudes are larger both between and within subspecies

Plumage

The plumage of the house sparrow is mostly different shades of grey and brown. The sexes exhibit strong **dimorphism**: the female is mostly **buffish** above and below, while the male has boldly coloured head markings, a reddish back, and grey underparts. The male has a dark grey **crown** from the top of its bill to its back, and chestnut brown flanking its crown on the sides of its head. It has black around its bill, on its throat, and on the spaces between its bill and eyes **(lores).** It has a small white stripe between the lores and crown and small white spots immediately behind the eyes (postoculars), with black patches below and above them. The underparts are pale grey or white, as are the cheeks, ear **coverts**, and stripes at the base of the head. The upper back and mantle are a warm brown, with broad black streaks, while the lower back, rump and uppertail coverts are greyish-brown.

The male is duller in fresh non-breeding plumage, with whitish tips on many feathers. Wear and preening expose many of the bright brown and black markings, including most of the black throat and chest patch, called the **"bib" or "badge".** The badge is variable in width and general size, and some scientists have suggested that patches signal social status or fitness. This hypothesis has led to a "veritable 'cottage industry'" of studies, which have only conclusively shown that patches increase in size with age. The male's bill is black in the breeding season and horn (dark grey) during the rest of the year.

The female has no black markings or grey crown. Its upperparts and head are brown with darker streaks around the mantle and a distinct pale **supercilium**. Its underparts are pale grey-brown. The female's bill is brownish-grey and becomes darker in breeding plumage approaching the black of the male's bill.

Juveniles are similar to the adult female but deeper brown below and paler above, with paler and less defined supercilia. Juveniles have broader buff feather edges, and tend to have looser, scruffier plumage, like moulting adults. Juvenile males tend to have darker throats and white postoculars like adult males, while juvenile female tend to have white throats. However, juveniles cannot be reliably sexed by plumage: some juvenile males lack any markings of the adult male, and some juvenile females have male features. The bills of young birds are light yellow to straw, paler than the female's bill. Immature males have paler versions of the adult male's markings, which can be very indistinct in fresh plumage. By their first breeding season, young birds generally are indistinguishable from other adults, though they may still be paler during their first year

Voice

Most house sparrow vocalisations are variations on its short and incessant chirping call. Transcribed as ***chirrup, tschilp***, or ***philip***, this note is made as a contact call by flocking or resting birds, or by males to proclaim nest ownership and invite pairing. In the breeding season the male gives this call repetitively, with emphasis and speed but not much rhythm, forming what is described either as a song or an "ecstatic call" similar to a song. Young birds also give a true song, especially in captivity, a warbling similar to that of the European greenfinch.

Aggressive males give a trilled version of their call, transcribed as **"chur-*chur-r-r-it-it-it-it*".** This call is also used by females in the breeding season, to establish dominance over males while displacing them to feed young or incubate eggs.[16]House sparrows give a nasal alarm call, the basic sound of which is transcribed as *quer*, and a shrill *chree* call in great distress. Another vocalisation is the "appeasement call", a soft *quee* given to inhibit aggression, usually given between birds of a mated pair. These vocalisations are not unique to the house sparrow, but are shared, with small variations, by all sparrows.

Distribution

The house sparrow originated in the Middle East and spread, along with agriculture, to most of Eurasia and parts of North Africa. Since the mid-nineteenth century, it has reached most of the world, chiefly due to deliberate introductions, but also through natural and shipborne dispersal. Its introduced range encompasses most of North America,Central America, southern South America, southern Africa, part of West Africa, Australia,New Zealand, and islands throughout the world. It has greatly extended its range in northern Eurasia since the 1850s, and continues to do so, as was shown by the colonisations around 1990 of Iceland and Rishiri Island, Japan. The extent of its range makes it the most widely distributed wild bird on the planet.

Habitat

The house sparrow is closely associated with human habitation and cultivation. It is not an obligate commensal of humans as some have suggested: Central Asian house sparrows usually breed away from humans in open country, and birds elsewhere are occasionally found away from humans. The only terrestrial habitats that the house sparrow does not inhabit are dense forest and tundra. Well adapted to living around humans, it frequently lives and even breeds indoors, especially in factories, warehouses and zoos. It has been recorded breeding in an English coal mine 640 m (2,100 ft) below ground, and feeding on the Empire State Building's observation deck at night. It reaches its greatest densities in urban centres, but its reproductive success is greater in suburbs, where insects are more abundant. On a larger scale, it is most abundant in wheat-growing areas such as the Midwestern United States.

It tolerates a variety of climates, but prefers drier conditions, especially in moist tropical climates. It has several adaptations to dry areas, including a high salt tolerance and an ability to survive without water by ingesting berries. In most of eastern Asia the house sparrow is entirely absent, replaced by the Eurasian tree sparrow. Where these two species overlap, the house sparrow is usually more common than the Eurasian tree sparrow, but one species may replace the other in a manner that ornithologist Maud Doria Haviland described as "random, or even capricious". In most of its range the house sparrow is extremely common, despite some declines, but in marginal habitats such as rainforest or mountain ranges, its distribution can be spotty

Social Behaviour

The house sparrow is a very social bird. It is gregarious at all seasons when feeding, often forming flocks with other types of bird. It roosts communally, its nests are usually grouped together in clumps, and it engages in social activities such as dust or water bathing and "social singing", in which birds call together in bushes. The house sparrow feeds mostly on the ground, but it flocks in trees and bushes. At feeding stations and nests, female house sparrows are dominant despite their smaller size, and in the reproductive period (usually spring or summer), being dominant, they can fight for males.

Sleep and Roosting

House sparrows sleep with the bill tucked underneath the scapular feathers. Outside of the reproductive season, they often roost communally in trees or shrubs. There is much communal chirping before and after the birds settle in the roost in the evening, as well as before the birds leave the roost in the morning. Some congregating sites separate from the roost may be visited by the birds prior to settling in for the night.

Body maintenance

Dust or water bathing is common and often occurs in groups. Anting is rare. Head scratching is done with the leg over the drooped wing

Feeding

As an adult, the house sparrow mostly feeds on the seeds of grains and weeds, but it is opportunistic and adaptable, and eats whatever foods are available. In towns and cities it often scavenges for food in garbage containers and congregates in the outdoors of restaurants and other eating establishments to feed on leftover food and crumbs. It can perform complex tasks to obtain food, such as opening automatic doors to enter supermarkets, clinging to hotel walls to watch vacationers on their balconies, and **nectar robbing kowhai flowers**. In common with many other birds, the house sparrow requires grit to digest the harder items in its diet. Grit can be either stone, often grains of masonry, or the shells of eggs or snails; oblong and rough grains are preferred.

Several studies of the house sparrow in temperate agricultural areas have found the proportion of seeds in its diet to be about 90 per cent. It will eat almost any seeds, but where it has a choice, it prefers oats and wheat. In urban areas, the house sparrow feeds largely on food provided directly or indirectly by humans, such as bread, though it prefers raw seeds. The house sparrow also eats some plant matter besides seeds, including buds, berries, and fruits such as grapes and cherries. In temperate areas, the house sparrow has an unusual habit of tearing flowers, especially yellow ones, in the spring.

Animals form another important part of the house sparrow's diet, chiefly insects, of which beetles, caterpillars, dipteran flies, and aphids are especially important. Various non-insect arthropods are eaten, as are molluscs and crustaceanswhere available, earthworms, and even vertebrates such as lizards and frogs. Young house sparrows are fed mostly on insects until about fifteen days after hatching. They are also given small quantities of seeds, spiders, and grit. In most places, grasshoppers and crickets are the most abundant foods of nestlings. True bugs, ants, sawflies, and beetles are also important, but house sparrows will take advantage of whatever foods are abundant to feed their young. House sparrows have been observed stealing prey from other birds, including American robins.

Locomotion

The house sparrow's flight is direct (not undulating) and flapping, averaging 45.5 kilometres per hour (28.3 miles per hour) and about 15 wingbeats per second. On the ground, the house sparrow typically hops rather than walks. It can swim when pressed to do so, by pursuit from predators. Captive birds have been recorded diving and swimming short distances underwater.

Dispersal and Migration

Most house sparrows do not move more than a few kilometres during their lifetime. However, there is limited migration in all regions. Some young birds disperse long distances, especially on coasts, and mountain birds move to lower elevations in winter. Two subspecies, *bactrianus* and *parkini*, are predominantly migratory. Unlike the birds in sedentary populations that migrate, birds of migratory subspecies prepare for migration by putting on weigh

Breeding

House sparrows can breed in the breeding season immediately following their hatching, and sometimes attempt to do so. Some birds breeding for the first time in tropical areas are only a few months old and still have juvenile plumage. Birds breeding for the first time are rarely successful in raising young, and reproductive success increases with age, as older birds breed earlier in the breeding season, and fledge more young. As the breeding season approaches, hormone releases trigger enormous increases in the size of the sexual organs and changes in day length lead males to start calling by nesting sites. The timing of mating and egg-laying varies geographically, and between specific locations and years. This is because a sufficient supply of insects is needed for egg formation and feeding nestlings.

Males take up nesting sites before the breeding season, by frequently calling beside them. Unmated males start nest construction and call particularly frequently to attract females. When a female approaches a male during this period, the male displays by moving up and down while drooping and shivering his wings, pushing up his head, raising and spreading his tail, and showing his bib. Males may try to mate with females while calling or displaying. In response, a female will adopt a threatening posture and attack a male before flying away, pursued by the male. The male displays in front of her, attracting other males, who also pursue and display to the female. This group display usually does not immediately result in copulations. Other males usually do not copulate with the female. Copulation is typically initiated by the female giving a soft *dee-dee-dee* call to the male. Birds of a pair copulate frequently until the female is laying eggs, and the male mounts the female repeatedly each time a pair mates.

The house sparrow is monogamous, and typically mates for life. Birds from pairs often engage in extra-pair copulations, so about 15 per cent of house sparrow fledglings are unrelated to their mother's mate. Male house sparrows guard their mates carefully to avoid being cuckolded, and most extra-pair copulation occurs away from nest sites. Males may sometimes have multiple mates, and bigamy is mostly limited by aggression between females. Many birds do not find a nest and a mate, and instead may serve as helpers around the nest for mated pairs, a role which increases the chances of being chosen to replace a lost mate. Lost mates of both sexes can be replaced quickly during the breeding season. The formation of a pair and the bond between the two birds is tied to the holding of a nest site, though paired house sparrows can recognise each other away from the nest

Nesting

Nest sites are varied, though cavities are preferred. Nests are most frequently built in the eaves and other crevices of houses. Holes in cliffs and banks, or tree hollows, are also used. A sparrow sometimes excavates its own nests in sandy banks or rotten branches, but more frequently uses the nests of other birds such as those of swallows in banks and cliffs, and old tree cavity nests. It usually uses deserted nests, though sometimes it usurps active ones. Tree hollows are more commonly used in North America than in Europe, putting the sparrows in competition with bluebirds and other North American cavity nesters, and thereby contributing to their population declines.

Especially in warmer areas, the house sparrow may build its nests in the open, on the branches of trees, especially evergreens and hawthorns, or in the nests of large birds such as storks or magpies. In open nesting sites, breeding success tends to be lower, since breeding begins late and the nest can easily be destroyed or damaged by storms. Less common nesting sites include street lights and neon signs, favoured for their warmth; and the old open-topped nests of other songbirds, which are then domed over.

The nest is usually domed, though it may lack a roof in enclosed sites. It has an outer layer of stems and roots, a middle layer of dead grass and leaves, and a lining of feathers, as well as of paper and other soft materials. Nests typically have external dimensions of 20 × 30 cm (8 × 12 in), but their size varies greatly. The building of the nest is initiated by the unmated male while displaying to females. The female assists in building, but is less active than the male. Some nest building occurs throughout the year, especially after moult in autumn. In colder areas house sparrows build specially created roost nests, or roost in street lights, to avoid losing heat during the winter. House sparrows do not hold territories, but they defend their nests aggressively against intruders of the same sex.

House sparrows' nests support a wide range of scavenging insects, including nest flies such as *Neottiophilum praestum*, *Protocalliphora* blowflies, and over 1,400 species of beetle.

Survival

In adult house sparrows, annual survival is 45–65 per cent.After fledging and leaving the care of their parents, young sparrows have a high mortality rate, which lessens as they grow older and more experienced. Only about 20–25 per cent of birds hatched survive to their first breeding season. The oldest known wild house sparrow lived for nearly two decades; it was found dead 19 years and 9 months after it was ringed in Denmark. The oldest recorded captive house sparrow lived for 23 years. The typical ratio of males to females in a population is uncertain due to problems in collecting data, but a very slight preponderance of males at all ages is usual

Predation

The house sparrow's main predators are cats and birds of prey, but many other animals prey on them, including corvids, squirrels, and even humans–the house sparrow has been consumed in the past by people in many parts of the world, and it still is in parts of the Mediterranean. Most species of bird of prey have been recorded preying on the house sparrow in places where there are extensive records. Accipiters and the merlin in particular are major predators, though cats are likely to have a greater impact on house sparrow populations. The house sparrow is also a common victim of roadkill; on European roads, it is the bird most frequently found dead

Parasites and Disease

The house sparrow is host to a huge number of parasites and diseases, and the effect of most is unknown. Ornithologist Ted R. Anderson listed thousands, noting that his list was incomplete. The commonly recorded bacterial pathogens of the house sparrow are often those common in humans, and include *Salmonella* and *Escherichia coli. Salmonella* is common in the house sparrow, and a comprehensive study of house sparrow disease found it in 13 per cent of sparrows tested.*Salmonella* epidemics in the spring and winter can kill large numbers of sparrows. The house sparrow hosts avian pox and avian malaria, which it has spread to the native forest birds of Hawaii. Many of the diseases hosted by the house sparrow are also present in humans and domestic animals, for which the house sparrow acts as a reservoir host. **Arboviruses** such as the West Nile virus, which most commonly infect insects and mammals, survive winters in temperate areas by going dormant in birds such as the house sparrow. There are a few records of disease extirpating house sparrow populations, especially from Scottish islands, but this seems to be rare.

The house sparrow is infested by a number of external parasites, which usually cause little harm to adult sparrows. In Europe, the most common mite found on sparrows is *Proctophyllodes*, the most common ticks are *Argas reflexus* and *Ixodes arboricola*, and the most common flea on the house sparrow is *Ceratophyllus gallinae*. A number of chewing lice occupy different niches on the house sparrow's body. *Menacanthus* lice occur across the house sparrow's body, where they feed on blood and feathers, while *Brueelia* lice feed on feathers and *Philopterus fringillae* occurs on the head.

Relationship with Humans

The house sparrow is closely associated with humans. They are believed to have become associated with humans around 10,000 years ago. Subspecies *bactrianus*is is least associated with humans and considered to be evolutionarily closer to the ancestral non-commensal populations. Usually, it is regarded as a pest, since it consumes agricultural products and spreads disease to humans and their domestic animals. Even birdwatchers often hold it in little regard because of its molestation of other birds. In most of the world the house sparrow is not protected by law. Attempts to control house sparrows include the trapping, poisoning, or shooting of adults; the destruction of their nests and eggs; or less directly, blocking nest holes and scaring off sparrows with noise, glue, or porcupine wire. However, the house sparrow can be beneficial to humans as well, especially by eating insect pests, and attempts at the large-scale control of the house sparrow have failed.

The house sparrow has long been used as a food item. From around 1560 to at least the nineteenth century in northern Europe, earthenware "sparrow pots" were hung from eaves to attract nesting birds so that the young could be readily harvested. Wild birds were trapped in nets in large numbers, and sparrow pie was a traditional dish, thought, because of the association of sparrows with lechery, to have aphrodisiac properties. Sparrows were also trapped as food forfalconers' birds and zoo animals. In the early part of the twentieth century, sparrow clubs culled many millions of birds and eggs in an attempt to control numbers of this perceived pest, but with only a localised impact on numbers. House sparrows have been kept as pets at many times in history, though they have no bright plumage or attractive songs, and raising them is difficult.

3. Indian Peafowl or Blue Peafowl (*Pavo cristatus* Linnaeus, 1758)

Classification

Kingdom: Animalia

Phylum: Chordata

Class: Aves

Order: Galliformes

Family: Phasianidae

Genus: *Pavo*

Species: *P. cristatus*

Binomial name: *Pavo cristatus* Linnaeus, 1758

The **Indian peafowl** or **blue peafowl** (*Pavo cristatus*), a large and brightly coloured bird, is a species of peafowl native to South Asia, but introduced in many other parts of the world.The male peafowl is predominantly blue with a fan-like crest of spatula-tipped wire-like feathers and is best known for the long train made up of elongated upper-tail covert feathers which bear colourful eyespots. These stiff feathers are raised into a fan and quivered in a display during courtship. Females lack the train, and have a greenish lower neck and duller brown plumage. The Indian peafowl lives mainly on the ground in open forest or on land under cultivation where they forage for berries, grains but also prey on snakes, lizards, and small rodents. Their loud calls make them easy to detect, and in forest areas often indicate the presence of a predator such as a tiger. They forage on the ground in small groups and usually try to escape on foot through undergrowth and avoid flying, though they fly into tall trees to roost.

The bird is celebrated in Indian and Greek mythology and is the national bird of India. The Indian peafowl is listed as of Least Concern by the **International Union for Conservation of Nature (IUCN).**

Description

Peacocks are a larger sized bird with a length from bill to tail of 100 to 115 cm (39 to 45 in) and to the end of a fully grown train as much as 195 to 225 cm (77 to 89 in) and weigh 4–6 kg (8.8–13.2 lb). The females, or peahens, are smaller at around 95 cm (37 in) in length and weigh 2.75–4 kg (6.1–8.8 lb). Indian peafowl are among the largest and heaviest representatives of the Phasianidae. Their size, colour and shape of crest make them unmistakable within their native distribution range. The male is metallic blue on the crown, the feathers of the head being short and curled. The fan-shaped crest on the head is made of feathers with bare black shafts and tipped with bluish-green webbing. A white stripe above the eye and a crescent shaped white patch below the eye are formed by bare white skin. The sides of the head have iridescent greenish blue feathers. The back has scaly bronze-green feathers with black and copper markings. The scapular and the wings are buff and barred in black, the primaries are chestnut and the secondaries are black. The tail is dark brown and the "train" is made up of elongated upper tail coverts (more than 200 feathers, the actual tail has only 20 feathers) and nearly all of these feathers end with an elaborate eye-spot. A few of the outer feathers lack the spot and end in a crescent shaped black tip. The underside is dark glossy green shading into blackish under the tail. The thighs are buff coloured. The male has a spur on the leg above the hind toe **(Plates 101-103).**

Plate 101: Male Neck Detail.

Plate 102: Peacock Dance.

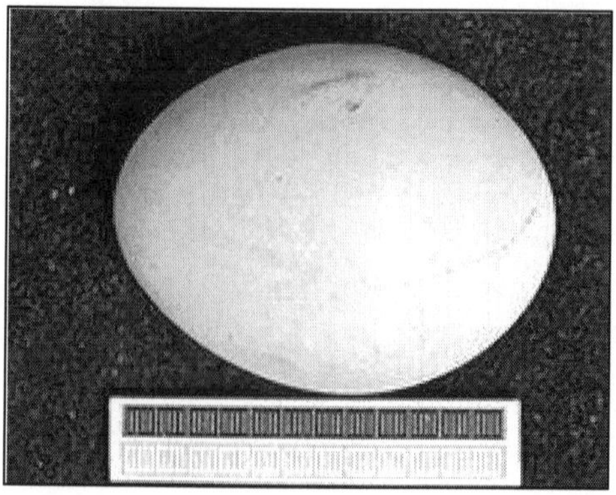

Plate 103: Egg.

The adult peahen has a rufous-brown head with a crest as in the male but the tips are chestnut edged with green. The upper body is brownish with pale mottling. The primaries, secondaries and tail are dark brown. The lower neck is metallic green and the breast feathers are dark brown glossed with green. The remaining underparts are whitish. Downy young are pale buff with a dark brown mark on the nape that connects with the eyes. Young males look like the females but the wings are chestnut coloured.

The most common calls are a loud *pia-ow* or *may-awe*. The frequency of calling increases before the Monsoon season and may be delivered in alarm or when disturbed by loud noises. In forests, their calls often indicate the presence of a predators such as the tiger. They also make many other calls such as a rapid series of *ka-aan.ka-aan* or a rapid *kok-kok*. They often emit an explosive low-pitched *honk!* When agitated.

Distribution and Habitat

The Indian peafowl is a resident breeder across the Indian subcontinent and is found in the drier lowland areas of Sri Lanka. In South Asia, it is found mainly below an altitude of 1,800 metres (1.1 mi) and in rare cases seen at about 2,000 metres (1.2 mi). It is found in moist and dry-deciduous forests, but can adapt to live in cultivated regions and around human habitations and is usually found where water is available. In many parts of northern India, they are protected by religious practices and will forage around villages and towns for scraps. Some have suggested that the peacock was introduced into Europe by Alexander the Great,]while others say the bird had reached Athens by 450 BCE and may have been introduced even earlier. It has since been introduced in many other parts of the world and has become feral in some areas.

Besides its native habitat, the bird has been introduced by humans to the United States, Mexico, Honduras, Colombia,Guyana, Suriname, Brazil, Uruguay, Argentina, South Africa, Madagascar, Mauritius, Réunion, Indonesia, Papua New Guinea and Australia, and elsewhere. In isolated cases, the Indian peafowl has been known to be able to adapt to harsher climates, such as those of northern Canada.

Behaviour and Ecology

Peafowl are best known for the male's extravagant display feathers which, despite actually growing from their back, are thought of as a tail. The "train" is in reality made up of the enormously elongated upper tail coverts. The tail itself is brown and short as in the peahen. The colours result not from any green or blue pigments but from the micro-structure of the feathers and the resulting optical phenomena. The long train feathers (and tarsal spurs) of the male develop only after the second year of life. Fully developed trains are found in birds older than four years. In northern India, these begin to develop each February and are moulted at the end of August. The moult of the flight feathers may be spread out across the year.

Peafowl forage on the ground in small groups, known as musters that usually have a cock and 3 to 5 hens. After the breeding season, the flocks tend to be made up only of females and young. They are found in the open early in the mornings and tend to stay in cover during the heat of the day. They are fond of dust-bathing and at dusk, groups walk in single file to a favourite waterhole to drink. When disturbed, they usually escape by running and rarely take to flight.

Peafowl produce loud calls especially in the breeding season. They may call at night when alarmed and neighbouring birds may call in a relay like series. Nearly seven different call variants have been identified in the peacocks apart from six alarm calls that are commonly produced by both sexes.

Peafowl roost in groups during the night on tall trees but may sometimes make use of rocks, buildings or pylons. In the Gir forest, they chose tall trees in steep river banks. Birds arrive at dusk and call frequently before taking their position on the roost trees. Due to this habit of congregating at the roost, many population studies are made at these sites. The population structure is not well understood. In a study in northern India (Jodhpur), the number of males was 170–210 for 100 females but a study involving evening counts at the roost site in southern India (Injar) suggested a ratio of 47 males for 100 females.

Breeding

Peacocks are **polygamous**, and the breeding season is spread out but appears to be dependent on the rains. Peafowls usually reach sexual maturity at the age of 2 to 3 years old. Several males may congregate at a lek site and these males are often closely related. Males at lek appear to maintain small territories next to each other and they allow females to visit them and make no attempt to guard harems. Females do not appear to favour specific males. The males display in courtship by raising the upper-tail coverts into an arched fan. The wings are held half open and drooped and it periodically vibrates the long feathers producing a ruffling sound. The cock faces the hen initially and struts and prances around and sometimes turns around to display the tail. Males may also freeze over food to invite a female in a form of courtship feeding. Males may display even in the absence of females. When a male is displaying, females do not appear to show any interest and usually continue their foraging. The peak season in southern India is April to May, January to March in Sri Lanka and June in northern India. The nest is a shallow scrape in the ground lined with leaves, sticks and other debris. Nests are sometimes placed on buildings and in earlier times have been recorded using the disused nest platforms of the white-rumped vultures. The clutch consists of 4–8 fawn to buff white eggs which are incubated only by the female. The eggs take about 28 days to hatch. The chicks are **nidifugous** and follow the mother around after hatching. Downy young may sometimes climb on their mothers' back and the female may carry them in flight to a safe tree branch. An unusual instance of a male incubating a clutch of eggs has been reported

Feeding

Peafowl are omnivorous and eat seeds, insects, fruits, small mammals and reptiles. They feed on small snakes but keep their distance from larger ones. In the Gir forest of Gujarat, a large percentage of their food is made up of the fallen berries of *Zizyphus*. Around cultivated areas, peafowl feed on a wide range of crops such as groundnut, tomato, paddy, chili and even bananas. Around human habitations, they feed on a variety of food scraps and even human excreta. In the countryside, it is particularly partial to crops and garden plants.

Mortality Factors

Adult peafowl can usually escape ground predators by flying into trees. Large animals such as leopards, dholes and tigers can sometimes ambush them however, and in some areas such as the Gir forest, peafowl are fairly common prey for such formidable predators. Foraging in groups provides some safety as there are more eyes to look out for predators. They are also sometimes hunted by large birds of prey such as the **crested hawk-eagle** and **rock eagle-owl**. Chicks are somewhat more prone to predation than adult birds. Adults living near human habitations are sometimes hunted by domestic dogs or by humans in some areas (southern Tamil Nadu) for folk remedies involving the use of "peacock oil".

In captivity, birds have been known to live for 23 years but it is estimated that they live for only about 15 years in the wild

4. Parakeets (Parrot)

Genaral species found in India:

(i) *Psittacula Krameri* – **main grainivorous**

(ii) *Psittacula cyanocephala*

(iii) *Psittacula himalayana*

Main Grainivorous (*Psittacula Krameri* Scopoli, 1769)

Classification

Kingdom: Animalia

Phylum: Chordata

Class: Aves

Order: Psittaciformes

Family: Psittaculidae

Genus: *Psittacula*

Species: *P. krameri* (Scopoli, 1769)

Binomial name: *Psittacula krameri*(Scopoli, 1769)

Introduction

Members of the parrot genus ***Psittacula*** or Afro-Asian ringnecked parakeets as they are commonly known in aviculture originates found from Africa to South-East Asia. It is a widespread group, with a clear concentration of species in south Asia, but also with representatives inAfrica and the islands of the Indian Ocean. This is the only genus of Parrot which has majority of its species in continental Asia. Of all the extant species only *Psittacula calthropae*, *Psittacula caniceps* and *Psittacula echo* do not have a representative subspecies in any part of mainland continental Asia. The rose-ringed parakeet, *Psittacula krameri*, is one of the most widely distributed of all parrots.

The **rose-ringed parakeet** (*Psittacula krameri*), also known as the **ring-necked parakeet**, is a gregarious tropical Afro-Asian parakeet species that has an extremely large range **(Plate 104)**.

Plate 104: Female on Left and Male on Right (*Psittacula krameri manillensis*).

The rose-ringed parakeet is sexually dimorphic. The adult male sports a red or black neck ring and the hen and immature birds of both sexes either show no neck rings, or display shadow-like pale to dark grey neck rings. Both sexes have a distinctive green colour. Rose-ringed parakeets measure on average 40 cm (16 in) in length, including the tail feathers, a large portion of their total length. Their average single-wing length is about 15–17.5 cm (5.9–6.9 in). In the wild, this is a noisy species with an unmistakable squawking call. It is herbivorous and not migratory.

One of the few parrot species that have successfully adapted to living in disturbed habitats; it has withstood the onslaught of urbanisation and deforestation. As a popular pet species, escaped birds have colonised a number of cities around the world. Since the population appears to be increasing, the species was evaluated as being of least concern by the IUCN in 2012, but its popularity as a pet and unpopularity with farmers have both reduced its numbers in some parts of its native range.

Distribution

Four subspecies are recognized, though they differ little:

- ☆ **African subspecies:**

  African rose-ringed parakeet (*P. k. krameri*): western Africa in Guinea,Senegal, and southern Mauritania, east to western Uganda and southern Sudan, Egypt. Resident among the Nile valley and certainly Giza, it is sometimes seen on the north coast and Sinai. The African parakeet also started to breed in Israel in the 1980s and is considered an invasive species.

  Abyssinian rose-ringed parakeet (*P. k. parvirostris*): northwest Somalia, west across northern Ethiopia to Sennar district, Sudan

- ☆ **Asian subspecies:**

  Indian rose-ringed parakeet (*P. k. manillensis*) originates from the southern Indian subcontinent and has feral and naturalized populations worldwide. In Australia, Great Britain (mainly around London), the United States, and other western countries, it is often referred to as the Indian ringneck parrot.

Boreal rose-ringed parakeet (*P. k. borealis*) is distributed in Bangladesh, Pakistan, northern India and Nepal to central Burma; introduced populations are found worldwide.

Diet

In the wild, rose-ringed parakeets usually feed on buds, fruits, vegetables, nuts, berries, and seeds. Wild flocks also fly several miles to forage in farmlands and orchards, causing extensive damage.

In India, they feed on cereal grains, and during winter also on pigeon peas. In Egypt during the spring, they feed on mulberry and in summer they feed on dates and nest inside palm trees and eat from sunflower and corn fields.

Reproduction

In north-west India, Indian rose-ringed parakeets form pairs from September to December. They do not have life mates and often breed with another partner during the following breeding season. During this cold season, they select and defend nest sites, thus avoiding competition for sites with other birds. Feeding on winter pea crops provides the female with nutrients necessary for egg production. From April to June, they care for their young. Fledgings are ready to leave the nest before monsoon.

Aviculture

Rose-ringed parakeets are popular as pets and they have a long history in aviculture. The ancient Greeks kept the Indian subspecies *P. krameri manillensis*, and the ancient Romans kept the African subspecies *P. krameri krameri*. Colour mutations of the Indian rose-ringed parakeet subspecies have become widely available in recent years.

Mimicry

Both males and females have the ability to mimic human speech. First, the bird listens to its surroundings, and then it copies the voice of the human speaker. Some people hand-raise rose-ringed parakeet chicks for this purpose. Such parrots then become quite tame and receptive to learning

5. Domestic Pigeon (*Columba livia domestica* Gmelin, 1789)

Classification

Kingdom: Animalia

Phylum: Chordata

Class: Aves

Order: Columbiformes

Family: Columbidae

Genus: *Columba*

Species: *C. livia*

Binomial name: *Columba livia domestica* Gmelin, 1789

Introduction

The **domestic pigeon** (*Columba livia domestica*) is a pigeon that was derived from the rock pigeon. The rock pigeon is the world's oldest domesticated bird.Mesopotamian cuneiform tablets mention the domestication of pigeons more than 5,000 years ago, as do Egyptian hieroglyphics. Research suggests that domestication of pigeons occurred as early as 10,000 years ago.

Pigeons have made contributions of considerable importance to humanity, especially in times of war. In war the homing ability of pigeons has been put to use by making them messengers. So-called war pigeons have carried many vital messages and some have been decorated for their services.

Reproduction

Domestic pigeons reproduce in a similar way to the wild rock pigeon. Generally humans will select breeding partners.Crop milk produced by parent birds may occasionally be replaced with artificial substitutes. Pigeons are extremely protective of their eggs, and in some cases will go to severe lengths to protect their productive eggs and have been known to seek revenge on those who interfere with their produtive process. Baby pigeons are called *squeakers* or*squabs*

Homing Pigeons

Trained domestic pigeons are able to return to the home loft if released at a location that they have never visited before and that may be up to **1000 km** away. A special breed, called homing pigeons has been developed through selective breeding to carry messages and members of this variety of pigeon are still being used in the sport of pigeon racing and the white release dove ceremony at weddings and funerals.

The ability a pigeon has to return home from a strange location necessitates two sorts of information. The first, called "map sense" is their geographic location. The second, "compass sense" is the bearing they need to fly from their new location in order to reach their home. Both of these senses, however, respond to a number of different cues in different situations. The most popular conception of how pigeons are able to do this is that they are able to sense the Earth's magnetic field with tiny magnetic tissues in their head (magnetoception). This is all the more surprising as they are not a migratory species, which is a fact used by some ornithologists to dispute this theory. Another theory is that pigeons have compass sense, which uses the position of the sun, along with an internal clock, to work out direction. However, studies have shown that if magnetic disruption or clock changes disrupt these senses, the pigeon can still manage to get home. The variability in the effects of manipulations to these senses of the pigeons indicates that there is more than one cue on which navigation is based and that map sense appears to rely on a comparison of available cues

6. House Crow (*Corvus splendens* Vieillot, 1817)

Classification

Kingdom: Animalia

Phylum: Chordata

Class: Aves

Order: Passeriformes

Family: Corvidae

Genus: *Corvus*

Species: *C. splendens*

Binomial name: *Corvus splendens* Vieillot, 1817

Introduction

The **house crow** (*Corvus splendens*), also known as the **Indian, greynecked, Ceylon** or **Colombo crow**, is a common bird of the crow family that is of Asian origin but now found in many parts of the world, where they arrived assisted by shipping. It is between the jackdaw and the carrion crow in size (40 cm (16 in) in length) but is slimmer than either. The forehead, crown, throat and upper breast are a richly glossed black, whilst the neck and breast are a lighter grey-brown in colour. The wings, tail and legs are black. There are regional variations in the thickness of the bill and the depth of colour in areas of the plumage **(Plate 105).**

Distribution and Habitat

It has a widespread distribution in southern Asia, being native to Nepal, Bangladesh, India, Pakistan, Sri Lanka, Maldivesand Laccadive Islands, South West Thailand and coastal southern Iran. It was introduced to East Africa around Zanzibar (about 1897) and Port Sudan. It arrived in Australia via ship but has up to now been exterminated. Recently, it has made its arrival in Europe and has been breeding in the Dutch harbour town Hook of Holland since 1998. An individual has been present in Cork Harbour on the south coast of Ireland since early September 2010.

Plate 105: House Crow.

Plate 106: Parents Feeding Nestlings.

Plate 107: Nest with Eggs.

It is associated with human settlements throughout its range, from small villages to large cities. In Singapore, there was a density of 190 birds/km^2 in 2001 with efforts to suppress the population in planning.

Due to a human population explosion in the areas it inhabits, this species has also proportionately multiplied. Being an omnivorous scavenger has enabled it to thrive in such circumstances.

The invasive potential for the species is great all over the tropics. This species is able to make use of resources with great flexibility and appears to be associated with humans, and no populations are known to exist independently of humans

Behaviour

Diet

House crows feed largely on refuse around human habitations, small reptilesand mammals, and other animals such as insects and other small invertebrates, eggs, nestlings, grain and fruits. House crows have also been observed swooping down from the air and snatching baby squirrels. Most food is taken from the ground, but also from trees as opportunity arises. They are highly opportunistic birds and given their omnivorous diet, they can survive on nearly anything that is edible. These birds can be seen near marketplaces and garbage dumps, foraging for scraps. They have also been observed to eat sand after feeding on carcass.

Nesting

At least some trees in the local environment seem to be necessary for successful breeding although house crows occasionally nest on telephone towers. It lays 3–5 eggs in a typical stick nest, and occasionally there are several nests in the same tree. InSouth Asia they are parasitized by the **Asian koel**. Peak breeding in India as well as Peninsular Malaysia is from April to July. Large trees with big crowns are preferred for nesting **(Plates 106-107).**

Roosting

House crows roost communally near human habitations and often over busy streets. A study in Singapore found that the preferred roost sites were in well-lit areas with a lot of human activity, close to food sources and in tall trees with dense crowns that were separated from other trees. The roost sites were often enclosed by tall buildings.

Voice

The voice is a harsh ***kaaa-kaaa.***

Crop Losses by Garinivorous Birds

Bajra: Parrot, *Passer domesticus.*

Wheat: Parrot including 13 species of birds. 41 per cent loss.

Paddy: Parrot including 39 species of birds. 6.5 per cent loss.

Sorghum: Parrot and Myna including 26 species of birds. 48 per cent loss.

Maize: Parrot including 10 species of birds. 20 per cent loss.

Sunflower: Parrot and crow. 10-90 per cent loss.

B. Frugivorous Bird Pests

Among 1200 bird species only 1 per cent birds causes damage on fruits which leads to poor quality and quantity and less earning. In India all the three species of parrot, Myna and crow are main frugivorous birds. Crow feeds on grape and cheeku, **Myna** on grape and anar, **Bulbul** on fig and ber and **Koel** on Mango.

Management

Birds are not allowed to kill as per rules.

1. Avoid
2. In the morning and evening make noise to make them away.
3. Hanging of predating birds.
4. Use of bright ribbons.
5. Covering of earheads by green leaves in Maize and sorghum.
6. Use of trap crop- chari.
7. Use of tobacco or neem extract as repellent or bio bird repellent.
8. Croppping in large area.
9. Creation of trouble to these birds.

Frugivorous Bats

1. *Rousettus leschenaultia*
2. *Pteropus giganteus*
3. *Cynopterus sphinx*

1. Leschenault's Rousette (*Rousettus leschenaultii* Desmarest, 1820)

Classification

Kingdom: Animalia

Phylum: Chordata

Class: Mammalia

Order: Chiroptera

Family: Pteropodidae

Genus: *Rousettus*

Species: *R. leschenaultii*

Binomial name: *Rousettus leschenaultii* Desmarest, 1820

Leschenault's rousette (*Rousettus leschenaultii*) is a species of fruit bat. The scientific name of the species is first published by Desmarest in 1820.

Description

R.leschenaultii is brown to grey-brown in colour with lighter underparts. It has long pale hairs under the chin, an elongated muzzle and large dark eyes. The head and body length measures 9.5–12 cm (3.7–4.7 in), the tail length measures 1–1.8 cm (0.39–0.71 in) and the forearm length is about 7.5–8.5 cm (3.0–3.3 in) **(Plate 108)**

Habit and Habitat

This bat species is found in a variety of habitats ranging form tropical forests to urban environments and it roosts in caves, old abandoned buildings and tunnels, and other such structures. A colonie of this animals can contain up to several thousand individuals. It feeds on fruits, nectar and flowers

Distribution

R.leschenaultii lives in Cambodja, Indonesia (Bali, Java and Sumatra), Laos, Peninsular Malaysia, Myanmar, Thailand, Vietnam, Bangladesh, China, India, Nepal, Pakistan and Sri Lanka.

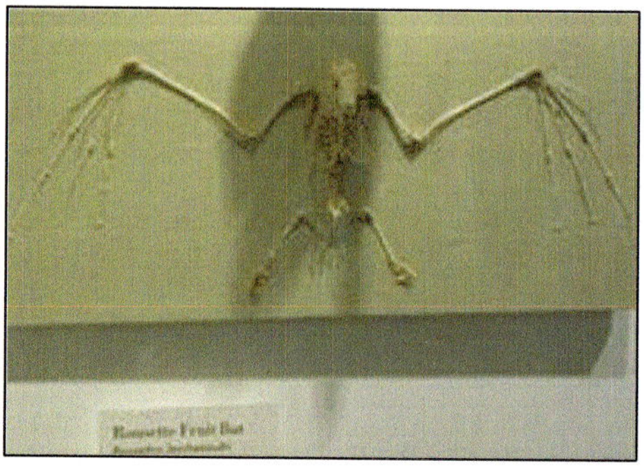

Plate 108: *Leschenault's rousette.*

2. Indian Flying Fox (*Pteropus giganteus* Brünnich, 1782)

Classification

Kingdom: Animalia

Phylum: Chordata

Class: Mammalia

Order: Chiroptera

Family: Pteropodidae

Genus: *Pteropus*

Species: *P. giganteus*

Binomial name: *Pteropus giganteus* (Brünnich, 1782)

The **Indian flying fox** (*Pteropus giganteus*), also known as the **greater Indian fruit bat**, is a species of flying fox in the family Pteropodidae. It is nocturnal and feeds mainly on ripe fruits, such as **mangoes** and **bananas**, and **nectar**. **(Plate 109)**

Distribution and Habitat

The Indian flying fox lives in tropical forests and swamps, near bodies of water. They also live on coastal islands, and it is theorized that ancestors migrated here long ago and found a liking to it, or were blown there and quickly forged a lifestyle. It

Plate 109: Indian Flying Fox.

is found in Bangladesh, China, India, the Maldives, Nepal, Pakistan, and Sri Lanka. It is classified as least concern due to its wide range and apparent adaptability

Feeding

The Indian flying fox is **frugivorous** or **nectarivorous**, *i.e.*, they eat fruits or lick nectar from flowers. At dusk, these bats forage for ripe fruit. While ingesting fruit, these bats expel waste that pollinates and disperse seeds.

Reproduction

Mating System: *P. giganteus* is a **polygynandrous** species, and breeds yearly from July to October. Births occur from February to May. Gestation period is typically 140 to 150 days. The average birth number is 1 to 2 pups. Among members of the genus *Pteropus*, pups are carried by the mother for the first few weeks of life, with weaning occurring around 5 months of age. Males do not participate in parental care. Young bats learn to fly at approximately 11 weeks of age. Reproductive maturity occurs at 1.5 years

Disease Transmission

Like other fruit bats, the Indian flying fox may be a natural reservoir for a number of diseases including members of the groups **Henipavirus** and **Coronavirus**. These can prove fatal to humans and domestic animals. Fruit bats are considered a delicacy by South Pacific Islanders as well as in Micronesia where, on the island of Guam, consumption has been suggested as a possible cause of **Lytico-Bodig disease.**

3. Greater Short-Nosed Fruit Bat or Short-Nosed Indian Fruit Bat (*Cynopterus sphinx* Vahl, 1797)

Classification

Kingdom: Animalia

Phylum: Chordata

Class: Mammalia

Order: Chiroptera

Family: Pteropodidae

Genus: *Cynopterus*

Species: *C. sphinx*

Binomial name: *Cynopterus sphinx* (Vahl, 1797)

The **greater short-nosed fruit bat** or **short-nosed Indian fruit bat** (*Cynopterus sphinx*) is a species of megabat in the family Pteropodidae. It is found in South and Southeast Asia, where it is known as **thala wawulaa** by Sinhalese people in Sri Lanka.

Description

These bats have a relatively long snout. Their upper parts are brown to grey-brown with paler under parts. The fur is very fine and silky. The ears and wing bones of *C. sphinx* are edged in white. Lower cheek teeth rounded without accessory cusps. The wing span of the adult is about 48 cm. Juveniles are lighter than adults. Average forearm length 70.2mm (64-79mm) **(Plate 110).**

Plate 110: Greater Short-nosed Fruit Bat.

Habitat

It is found in Pakistan, Bangladesh, Bhutan, Cambodia, China,India, Indonesia, Laos, Malaysia, Myanmar, Philippines, Sri Lanka, Thailand, and Vietnam. *C. sphinx* is common in tropical forests and areas where fruit crops are cultivated. They can also be found in grassland and mangrove forests. They typically nest high in palm trees. The bats chew the fronds of the palms to construct fairly simple tents. These bats are also known to construct tents by closely inter weaving the leaves and twigs of creeping vines which cover buildings, but such nests are constructed only when palms are not available

Behaviour and Breeding

The species is gregarious, and typically roosts in same sex groups of 8-9 individuals. The sexes remain separate until the mating season, when group size increases. They are polygynous and it is usual for 6-10 males and 10-15 females to share palm frond tents during the breeding season. *C. sphinx* is the only non-primate species known to show **fellatio**, which enhances copulation time in the species. **Copulation** by males is dorsoventral and the females lick the shaft or the base of the male's penis, but not the glans which has alreadypenetrated the vagina. While the females do this, the penis is not withdrawn and research has shown a positive relationship between length of the time that the penis is licked and the duration of copulation. Post copulation genital grooming has also been observed. Males stay with females for some time after mating, but later return to same sex groups.

The adult sex ratio is very female biased. Researchers attribute this to the relatively rapid maturation of females compared to males. In Central India, *C. sphinx* breeds twice per year. Females produce a single young at a time. Each half of the bicornate uterus functions during alternate breeding cycles. The first pregnancy cycle occurs from October through February/March. Mating occurs immediately **postpartum,** and a second offspring is born in July. Gestation period is about 3–5 months. In 72 per cent of bats, the first pregnancy occurs in the right horn of the uterus. The **corpus luteum** in the right ovary persists for some time after the pregnancy and prevents ovulation from occurring in the right ovary during the second breeding cycle. This creates the pattern of alternate functioning of the two horns of the uterus. However, the corpus luteum in the left ovary does not persist until the beginning of the next breeding cycle. As yet, no reason has been found for the dominance of the right horn during the first breeding cycle. Newborn bats weigh about 13.5 g and have a wingspan of 24 cm. By the time of weaning at 4 weeks of age, young bats weigh 25 g and have wings spanning 36 cm. Female short-nosed fruit bats reach sexual maturity at 5–6 months of age, but males are not capable of breeding until they are a year old.

These bats are frugivorous, locate their preferred food items by scent. They have been described as voracious feeders, eating more than their body weight in food in one sitting. Some preferred fruits include ripe guava, banana, chikoo, dates and lychees.

Short-nosed fruit bats inflict serious damage on many fruit crops, and are considered pests. In addition, these bats are possible vectors for **Japanese encephalitis,** which is serious disease in humans. These bats are important dispersers of date palm seeds, and pollinate many night blooming flowers.

Frugivory, nectarivory and folivory are well understood in *C. sphinx;* in addition, geophagy behaviour has also been reported in this species recently, and is suggested to represent an 'adaptive behavioural plasticity' in the foraging behaviour of the *C.sphinx*.

Management

1. Covering of fuits in pouch
2. Lighting and sound in the night
3. Web around the orchard
4. Timely picking of ripening fruits

Chapter 7
Monkeys

Three species of monkey are found in India:

1. *Macaca mulatta*
2. *Macaca radiate*
3. *Semnopithecus entellus*

1. Rhesus Macaque (*Macaca mulatta* Zimmermann, 1780)

Classification

Kingdom: Animalia

Phylum: Chordata

Class: Mammalia

Order: Primates

Family: Cercopithecidae

Genus: *Macaca*

Species: M. mulatta

Binomial name: *Macaca mulatta* (Zimmermann, 1780)

The **rhesus macaque** (*Macaca mulatta*) is one of the best-known species of **Old World monkeys**. It is listed as least concern in the IUCN Red List of Threatened Species in view of its wide distribution, presumed large population, and its tolerance of a broad range of habitats. Native to South, Central, and Southeast Asia, troops of *M. mulatta* inhabit a great variety of habitats, from grasslands to arid and forested areas, but also close to human settlements.

Characteristics

The rhesus macaque is brown or grey in color and has a pink face, which is bereft of fur. Its tail is of medium length and averages between 20.7 and 22.9 cm (8.1 and 9.0 in). Adult males measure about 53 cm (21 in) on average and weigh about 7.7 kg (17 lb). Females are smaller, averaging 47 cm (19 in) in length and 5.3 kg (12 lb) in weight. Rhesus macaques have, on average, 50 vertebrae. Their ratio of arm length to leg length is 89 per cent. They have dorsal scapulae and a wide rib cage **(Plate 111).**

Plate 111: Rhesus Macaque.

The rhesus macaque has 32 teeth with a dental formula of 2.1.2.3/2.1.2.3 and bilophodont molars. The upper molars have four cusps: paracone, metacone, protocone, and hypocone. The lower molars also have four cusps: metaconid, protoconid, hypoconid, and entoconid.

Distribution and Habitat

Rhesus macaques are native to northern India, Bangladesh, Pakistan, Nepal, Burma, Thailand, Afghanistan, Vietnam, southern China, and some neighboring areas. They have the widest geographic ranges of any nonhuman primate, occupying a great diversity of altitudes throughout Central, South, and Southeast Asia. Inhabiting arid, open areas, rhesus macaques may be found in grasslands, woodlands, and in mountainous regions up to 2,500 m (8,200 ft) in elevation. They are regular swimmers. Babies as young as a few days old can swim, and adults are known to swim over a half mile between islands, but are often found drowned in small groups where their drinking waters lie. Rhesus macaques are noted for their tendency to move from rural to urban areas, coming to rely on handouts or refuse from humans. They adapt well to human presence, and form larger troops in human-dominated landscapes than in forests.

The southern and the northern distributional limits for rhesus and bonnet macaques, respectively, currently run parallel to each other in the western part of India, are separated by a large gap in the center, and converge on the eastern coast of the peninsula to form a distribution overlap zone. This overlap region is characterized by the presence of mixed-species troops, with pure troops of both species sometimes occurring even in close proximity to one another. The range extension of rhesus macaque – a natural process in some areas, and a direct consequence of introduction by humans in other regions – poses grave implications for the endemic and declining populations of bonnet macaques in southern India

Ecology and Behaviour

Rhesus macaques are **diurnal animals**, and both **arboreal** and **terrestrial**. They are quadrupedal and, when on the ground, they walk digitigrade and plantigrade. They are mostly herbivorous, feeding mainly on fruit, but also eating seeds, roots,buds, bark, and cereals. They are estimated to consume around 99 different plant species in 46 families. During the monsoon season, they get much of their water from ripe and succulent fruit. Macaques living far from water sources lick dewdrops from leaves and drink rainwater accumulated in tree hollows. They have also been observed eating termites, grasshoppers, ants, and beetles. When food is abundant, they are distributed in patches, and forage throughout the day in their home ranges. They drink water when foraging, and gather around streams and rivers. Rhesus macaques have specialized pouch-like cheeks, allowing them to temporarily hoard their food.

In psychological research, rhesus macaques have demonstrated a variety of complex cognitive abilities, including the ability to make same-different judgments, understand simple rules, and monitor their own mental states. They have even been shown to demonstrate self-agency, an important type of self-awareness. In 2014, onlookers at a train station in Kanpur, India, documented a rhesus monkey, knocked unconscious by overhead power lines that was revived by another rhesus that systematically administered a series of resuscitative actions.

Group Structure

Like other macaques, rhesus troops comprise a mixture of 20–200 males and females. Females may outnumber the males by a ratio of 4:1. Males and females both have separate hierarchies. Female **philopatry**, common among social mammals, has been extensively studied in rhesus macaques. Females tend not to leave the social group, and have highly stable **matrilineal** hierarchies in which a female's rank is dependent on the rank of her mother. In addition, a single group may have multiple matrilineal lines existing in a hierarchy, and a female outranks any unrelated females that rank lower than her mother. Rhesus macaques are unusual in that the youngest females tend to outrank their older sisters. This is likely because young females are more fit and fertile. Mothers seem to prevent the older daughters from forming coalitions against her. The youngest daughter is the most dependent on the mother, and would have nothing to gain from helping her siblings in overthrowing their mother. Since each daughter had a high rank in her early years, rebelling against her mother is discouraged. Juvenile male macaques also exist in matrilineal lines, but once they reach four to five years of age, they are driven out of their natal groups by the dominant male. Thus, adult males gain dominance by age and experience.

In the group, macaques position themselves based on rank. The "**central male subgroup**" contains the two or three oldest and most dominant males which are codominant, along with females, their infants, and juveniles. This subgroup occupies the center of the group and determines the movements, foraging, and other routines. The females of this subgroup are also the most dominant of the entire group. The farther to the periphery a subgroup is, the less dominant it is. Subgroups on the periphery of the central group are run by one dominant male, of a rank lower than the central males, and he maintains order in the group, and communicates messages between the central and peripheral males. A subgroup of subordinate, often subadult, males occupy the very edge of the groups, and have the responsibility of communicating with other macaque groups and making alarm calls. Rhesus social behaviour has been described as despotic, in that high-ranking individuals show little tolerance and frequent and often severe aggression towards non-kin

Communication

Rhesus macaques interact using a variety of facial expressions, vocalizations, body postures, and gestures. Perhaps the most common facial expression the macaque makes is the "silent bared teeth" face. This is made between individuals of different social ranks, with the lower-ranking one giving the expression to its superior. A less-dominant individual also makes a "fear grimace", accompanied by a scream, to appease or redirect aggression. Another submissive behavior is the "present rump", where an individual raises its tail and exposes its genitals to the dominant one. A dominant individual threatens another individual by standing quadrupedally and making a silent "open mouth stare" accompanied by the tail sticking straight. During movements, macaques make coos and grunts. These are also made during affiliative interactions, and approaches before grooming. When they find rare food of high quality, macaques emit warbles, harmonic arches, or chirps. When in threatening situations, macaques emit a single loud, high-pitched sound called a shrill bark. Screeches, screams, squeaks, pant-threats, growls, and barks are used during aggressive interactions. Infants "**gecker**" to attract their mother's attention

Reproduction

Adult male macaques try to maximize their reproductive success by entering into sex with females, both in and outside the breeding period. Females prefer to mate with males that will increase the survival of their young. Thus, a consort male provides resources for his female and protects her from predators. Larger, more dominant males are more likely to provide for the females. The breeding period can last up to 11 days, and a female usually mates with four males during that time. Male rhesus macaques have not been observed to fight for access to sexually receptive females, although they suffer more wounds during the mating season. Female macaques first breed when they are four years old, and reach **menopause** at around 25 years of age. When mating, a male rhesus monkey usually ejaculates less than 15 seconds after sexual penetration. Male macaques generally play no role in raising the young, but do have peaceful relationships with the offspring of their consort pairs.

Manson and Parryfound those free-ranging rhesus macaques avoid inbreeding. Adult females were never observed to copulate with males of their own matrilineage during their fertile periods.

Mothers with one or more immature daughters in addition to their infants are in contact with their infants less than those with no older immature daughters, because the mothers may pass the parenting responsibilities to their daughters. High-ranking mothers with older immature daughters also reject their infants significantly more than those without older daughters, and tend to begin mating earlier in the mating season than expected based on their dates of parturition the preceding birth season. Infants farther from the center of the groups are more vulnerable to infanticide from outside groups. Some mothers abuse their infants, which is believed to be the result of controlling parenting styles

2. Bonnet Macaque (*Macaca radiate* É. Geoffroy, 1812)

Classification

Kingdom: Animalia

Phylum: Chordata

Class: Mammalia

Order: Primates

Family: Cercopithecidae

Genus: *Macaca*

Species: *M. radiata*

Binomial name: *Macaca radiata* (É. Geoffroy, 1812)

The **bonnet macaque** (*Macaca radiata*) is a macaque endemic to southern India. Its distribution is limited by the Indian Ocean on three sides and the Godavari and Tapti Rivers along with a related competing species of rhesus macaque in the north. Land use changes in the last few decades have resulted in changes in its distribution boundaries with the rhesus macaque, raising concern for its status in the wild.

This Old World monkey is a diurnal animal. It is 35–60 cm long plus a tail of 35–68 cm. Males weigh 5.5 to 9 kg, females 3.5 to 4.5 kg. Captive monkey can live up to 35 years **(Plate 112)**.

The bonnet macaque feeds on fruits, nuts, seeds, flowers, invertebrates and cereals. In southern India, this macaque exists as **commensal** to humans, feeding on food given by humans and raiding crops and houses.

Two subspecies of bonnet macaques have been identified: *M. r. radiata* and*M. r. diluta*.

Plate 112: Bonnet Macaque.

Behaviour

The bonnet macaque has a very wide range of gestures and behaviors, which can be easily differentiated. 'Lip smacking' is one of the most common affiliative behaviors, where one individual may open and close its mouth in rapid succession, with its tongue between its teeth and its lips pressing against each other, giving an audible sound. A 'grimace' is the most common gesture of fear or submission that a subordinate shows to a dominant individual during aggressive encounters. It consists of pulling back its upper lip, showing its upper teeth. They also have distinct alarm calls for predators such as python and leopard.

Social Structure

The bonnet macaque, like other macaques, shares a linear dominance hierarchy; the 'alpha' male is the most dominant male of the troop, followed by a 'beta' male and a 'gamma' male, and so on according to their dominance. Similarly, females also follow this linear herarchy. The male and female hierarchies are different and of a nonoverlapping or nonmixing types. Males are usually dominant over females.

The females' dominance hierarchy is stable (rarely changes), whilst the males' dominance hierarchy is very dynamic. In the male hierarchy, males close in rank often fight to rise in rank. A male has the best chance of obtaining high rank in his prime age, resulting in the greatest benefits to reproduction. High-ranking individuals have first access to breeding females. Females are receptive during only a few months in a year, resulting in competition between males. In this situation, the ranks established by aggressive encounters come into play. Most of these aggressive encounters are easily resolved, but competition between similarly built or similarly aggressive males results in brutal and sometime fatal fights. Different males may employ various means to rise in rank. Coalition formation between unrelated males to oust a more dominant male has been observed. Males often move from troop to troop to gain higher rank with the resulting benefits. However, males remaining in a single troop have been observed to rise to become dominant male of that troop.

In the case of females, the stable dominance hierarchy is a result of female **philopatry**, when individuals tend to remain with the troop into which they are born. This results in the formation of **matrilinear** groupings of closely related females. These matrilines help each other during antagonistic interactions. As the matrilines are continuously reinforced with new births of females in a troop, ranks rarely change. A few rare cases of rank reversal for females have been recorded in which matrilines have become extremely depleted due to few female births. Male infants would not help to reinforce the matrilines, as they move off to new troops leaving their natal troops

3. Gray Langurs or Hanuman Langurs (*Semnopithecus entellus* Dufresne, 1797)

Classification

Kingdom: Animalia

Phylum: Chordata

Class: Mammalia

Order: Primates

Family: Cercopithecidae

Genus: *Semnopithecus* Desmarest, 1822

Species: *Entellus* Dufresne, 1797

Binomial name: *Semnopithecus entellus* Dufresne, 1797

Gray langurs or **Hanuman langurs**, the most widespread langurs of South Asia, are a group of Old World monkeys constituting the entirety of the genus***Semnopithecus***. All taxa have traditionally been placed in the single species *Semnopithecus entellus*. In 2001, it was recommended that several distinctive former subspecies should be given species status, so that seven species are recognized. A taxonomic classification with fewer species has also been proposed. Genetic evidence suggests that the Nilgiri langur and purple-faced langur, which usually are placed in the genus *Trachypithecus*, actually belong in *Semnopithecus*.

Gray langurs are large and fairly terrestrial, inhabiting forest, open lightly wooded habitats, and urban areas on the Indian subcontinent. Most species are found at low to moderate altitudes, but the Nepal gray langur and Kashmir gray langur occur up to 4,000 metres (13,000 ft) in the Himalayas.

Characteristics

These langurs are largely gray (some more yellowish), with a black face and ears. Externally, the various species mainly differ in the darkness of the hands and feet, the overall color and the presence or absence of a crest. Typically all north Indian gray langurs have their tail tips looping towards their head during a casual walk whereas all south Indian and Sri Lankan gray langurs have an inverted "U" shape or a "S" tail carriage pattern. There are also significant variations in the size depending on the sex, with the male always larger than the female. The head-and-body length is from 51 to 79 cm (20 to 31 in). Their tails, at 69 to 102 cm (27 to 40 in) are always longer than their bodies. Langurs from the southern part of their range are smaller than those from the north. At 26.5 kg (58 lb), the heaviest langur ever recorded was a male Nepal gray langur. The larger gray langurs are rivals for the largest species of monkey found in Asia. The average weight of gray langurs is 18 kg (40 lb) in the males and 11 kg (24 lb) in the females.

Langurs mostly walk quadrupedally and spend half their time on the ground and the other half in the trees. They will also make bipedal hops, climbing and descending supports with the body upright, and leaps. Langurs can leap 3.7–4.6 m (12–15 ft) horizontally and 10.7–12.2 m (35–40 ft) in descending **(Plate 113).**

Plate 113: Gray Langur.

Distribution and Habitat

The entire distribution of all gray langur species stretches from the Himalayas in the north to Sri Lanka in the south, and from Bangladesh in the east to Pakistan in the west. They possibly occur in Afghanistan. The bulk of the gray langur distribution is within India, and all seven currently recognized species have at least a part of their range in this country.

Gray langurs can adapt to a variety of habitats. They inhabit arid habitats like deserts, tropical habitats like tropical rainforests and temperate habitats like coniferous forests, deciduous habitats and mountains habitats. They are found at sea level to altitudes up to 4,000 m (13,000 ft). They can adapt well to human settlements, and are found in villages, towns and areas with housing or agriculture. They live in densely populated cities like Jodhpur, which has a population numbering up to a million.

Ecology and Behaviour

Gray langurs are **diurnal**. They sleep during the night in trees but also on man-made structures like towers and electric poles when in human settlements. When resting in trees, they generally prefer the highest branches.

Ungulates like **bovine** and **deer** eat food dropped by foraging langurs. Langurs are preyed upon by leopards,dholes and tigers. Wolves, jackals and pythons may also prey on langurs.

Diet

Gray langurs are primarily **herbivores**. However, unlike some other colobines they do not depend on leaves and leaf buds of herbs, but will also eat **coniferous** needles and cones, fruits and fruit buds, evergreen petioles, shoots and roots, seeds, grass, bamboo, fern rhizomes, mosses, and lichens. Leaves of trees and shrubs rank at the top of preferred food, followed by herbs and grasses. Non-plant material consumed include spider webs, termite mounds and insect larvae. They forage on agricultural crops and other human foods, and even accept handouts. Although they occasionally drink, langurs get most of their water from the moisture in their food

Social Structure

Gray langurs exist in three types of groups:

- One-male groups, comprising one adult male, several females and offspring;
- Multiple-male groups, comprising males and females of all ages;
- All-male groups.

All-male groups tend to be the smallest of the groups and can consist of adults, subadults, and juveniles. Some populations have only multiple-male groups as mixed sex groups, while others have only one-male groups as mixed sexed groups.

Some evidence suggests multiple-male groups are temporary and exist only after a takeover, and subsequently split into one-male and all-male groups.

Social hierarchies exist for all group types. In all-male groups, dominance is attained through aggression and mating success. With sexually mature females, rank is based on physical condition and age. The younger the female, the higher the rank. Dominance rituals are most common among high-ranking langurs. Most changes in social rank in males take place during changes in group members. An adult male may remain in a one-male group for 45 months. The rate of male replacement can occur quickly or slowly depending on the group.

Females within a group are matrilineally related. Female memberships are also stable, but less so in larger groups. Relationships between the females tend to be friendly. They will do various activities with each together, such as foraging, traveling and resting. They will also groom each other regardless of their rank. However, higher-ranking females give out and receive grooming the most. In addition, females groom males more often than the other way around. Male and female relationships are usually positive. Relationships between males can range from peaceful to violent. While females remain in their natal groups, males will leave when they reach adulthood.[26] Relationships between groups tend to be hostile. High-ranking males from different groups will display, vocalize, and fight among themselves.

Reproduction and Parenting

In one-male groups, the resident male is usually the sole breeder of the females and sires all the young. In multiple-male groups, the highest-ranking male fathers most of the offspring, followed by the next-ranking males and even outside males will father young. Higher-ranking females are more reproductively successful than lower-ranking ones.

Female gray langurs do not make it obvious that they are in estrous. However, males are still somehow able to deduce the reproduction state of females. Females signal that they are ready to mate by shuddering the head, lowering the tail, and presenting their anogenital regions. Such solicitations do not always lead to copulation. When langurs mate, they are sometimes disrupted by other group members. Females have even been recorded mounting other females.

The gestation period of gray langur lasts around 200 days, at least at Jodhpur, India. In some areas, reproduction is year-around. Year-round reproduction appears to occur in populations that capitalize on human-made foods. Other populations have seasonal reproduction.

Infanticide is common among gray langurs. Most infanticidal langurs are males that have recently immigrated to a group and driven out the prior male. These males only kill infants that are not their own. Infanticide is more commonly reported in one-male groups, perhaps because one male monopolizing matings drives the evolution of this trait. In multiple-male groups, the costs for infanticidal males are likely to be high as the other males may protect the infants and they can't ensure that they'll sire young with other males around. Nevertheless, infanticide does occur in these groups, and is suggested that such practices serve to return a female to estrous and gain the opportunity to mate.

Females usually give birth to a single infant, although twins do occur. Most births occur during the night. Infants are born with thin, dark brown or black hair and pale skin. Infants spend their first week attach themselves to their mothers' chests and mostly just suckle or sleep. They do not move much in terms of locomotion for the first two weeks of their life. As they approach their sixth week of life, infants vocalize more. They use squeaks and shrieks to communicate stress. In the following months, the infants are capable of quadrupedal locomotion and can walk, run and jump by the second and third months. Alloparenting occurs among langurs, starting when the infants reach two years of age. The infant will be given to the other females of the group. However, if the mother dies, the infant usually follows. Langurs are weaned by 13 months.

Vocalizations

Gray langurs are recorded to make a number of vocalizations:

- ☆ Loud calls or whoops made only by adult males during displays;
- ☆ Harsh barks made by adult and subadult males when surprised by a predator;
- ☆ Cough barks made by adults and subadults during group movements;
- ☆ Grunt barks made mostly by adult males during group movements and agonistic interactions;
- ☆ Rumble screams made in agonistic interactions;
- ☆ Pant barks made with loud calls when groups are interacting;
- ☆ Grunts made in many different situations, usually in agonistic ones;
- ☆ Honks made by adult males when groups are interacting;
- ☆ Rumbles made during approaches, embraces, and mounts;
- ☆ Hiccups made by most members of a group when they find another group.

Chapter 8
The Wild Boar/Wild Swine/ Eurasian Wild Pig (*Sus scrofa* Linnaeus, 1758)

Classification

Kingdom: Animalia

Phylum: Chordata

Class: Mammalia

Order: Artiodactyla

Family: Suidae

Genus: *Sus*

Species: S. scrofa

Binomial name: *Sus scrofa* Linnaeus, 1758

The **wild boar** (*Sus scrofa*), also known as the **wild swine** or **Eurasian wild pig** is a suid native to much of Eurasia, North Africa, and the Greater Sunda Islands. Human intervention has spread its range further, making the species one of the widest-ranging mammals in the world, as well as the most widely spread suiform. Its wide range, high numbers, and adaptability mean that it is classed as least concern by the IUCN. The animal probably originated in Southeast Asia during the Early Pleistocene, and outcompeted other suid species as it spread throughout the Old World.

As of 2005, up to 16 subspecies are recognised, which are divided into four regional groupings based on skull height and lacrimal bone length. The species lives

Terminology

Designation	Age	Image
Squeaker	0–10 months	
Juvenile	10–12 months	
Pig of the sounder **Boar of the 4th/5th/6th year**	Two years 3–5 years	
Old boar **Grand old boar**	Six years Over seven years	

Plate 114: Different Stages of Wild Boar.

in matriarchal societies consisting of interrelated females and their young (both male and female). Fully grown males are usually solitary outside of the breeding season. The grey wolf is the wild boar's main predator throughout most of its range except in the Far East and the Lesser Sunda Islands, where it is replaced by the tiger and Komodo dragonrespectively. It has a long history of association with humans, having been the ancestor of most domestic pig breeds and a big-game animal for millennia.

Description

The wild boar is a bulky, massively built suid with short and relatively thin legs. The trunk is short and massive, while the hindquarters are comparatively underdeveloped. The region behind the shoulder blades rises into a hump, and the neck is short and thick, to the point of being nearly immobile. The animal's head is very large, taking up to one third of the body's entire length. The structure of the head is well suited for digging. The head acts as a plow, while the powerful neck muscles allow the animal to upturn considerable amounts of soil: it is capable of digging 8–10 cm (3.1–3.9 in) into frozen ground and can upturn rocks weighing 40–50 kg (88–110 lb). The eyes are small and deep-set, and the ears long and broad. The species has well developed canine teeth, which protrude from the mouths of adult males. The middle hooves are larger and more elongated than the lateral ones, and are capable of quick movements. The animal can run at a maximum speed of 40 km/h and jump at a height of 140–150 cm (55–59 in). Sexual dimorphism is very pronounced in the species, with males being typically 5-10 per cent larger and 20-30 per cent heavier than females. Males also sport a mane running down the back, which is particularly apparent during autumn and winter. The canine teeth are also much more prominent in males, and grow throughout life. The upper canines are relatively short and grow sideways early in life, though gradually curve upwards. The lower canines are much sharper and longer, with the exposed parts measuring 10–12 cm (3.9–4.7 in) in length. In the breeding period, males develop a coating of subcutaneous tissue, which may be 2–3 cm (0.79–1.18 in) thick, extending from the shoulder blades to the rump, thus protecting vital organs during fights. Males sport a roughly egg-sized sack near the opening of the penis, which collects urine and emits a sharp odour. The purpose of this is not fully understood **(Plate 114).**

Adult size and weight is largely determined by environmental factors; boars living in arid areas with little productivity tend to attain smaller sizes than their counterparts inhabiting areas with abundant food and water. In most of Europe, males average 75–100 kg (165–220 lb) in weight, 75–80 cm (30–31 in) in shoulder height and 150 cm (59 in) in body length, whereas females average 60–80 kg (130–180 lb) in weight, 70 cm (28 in) in shoulder height and 140 cm (55 in) in body length. In Europe's Mediterranean regions, males may reach average weights as low as 50 kg (110 lb) and females 45 kg (99 lb), with shoulder heights of 63–65 cm (25–26 in). In the more productive areas of Eastern Europe, males average 110–130 kg (240–290 lb) in weight, 95 cm (37 in) in shoulder height and 160 cm (63 in) in body length, while females weigh 95 kg (209 lb), reach 85–90 cm (33–35 in) in shoulder height and 145 cm (57 in) in body length. In Westernand Central Europe, the largest males weigh 200 kg (440 lb) and females 120 kg (260 lb). In Eastern Europe, large males can reach brown bear-like sizes, weighing 270 kg (600 lb) and measuring 110–118 cm (43–46 in) in shoulder height. Some adult males in Ussuriland and Manchuria have been recorded to weigh 300–350 kg (660–770 lb) and measure 125 cm (49 in) in shoulder height. Adults of this size are generally immune from wolf predation. Such giants are rare in modern times, due to past overhunting preventing animals from attaining their full growth.

The winter coat consists of long, coarse bristles underlaid with short brown downy fur. The length of these bristles varies along the body, with the shortest being around the face and limbs and the longest running along the back. These back bristles form the aforementioned mane prominent in males, and stand erect when the animal is agitated. Colour is highly variable; specimens around Lake Balkhash are very lightly coloured, and can even be white, while some boars fromBelarus and Ussuriland can be black. Some subspecies sport a light coloured patch running backwards from the corners of the mouth. Coat colour also varies with age, with piglets having light brown or rusty-brown fur with pale bands extending from the flanks and back.

The wild boar produces a number of different sounds which are divided into three categories:

- **Contact calls**: Grunting noises which differ in intensity according to the situation. Adult males are usually silent, while females frequently grunt and piglets whine. When feeding, boars express their contentment through purring. Studies have shown that piglets imitate the sounds of their mother, thus different litters may have unique vocalisations.
- **Alarm calls**: Warning cries emitted in response to threats. When frightened, boars make loud huffing *ukh! ukh!*sounds or emit screeches transcribed as *gu-gu-gu*.
- **Combat calls**: High-pitched, piercing cries.

Its sense of smell is very well developed, to the point that the animal is used for drug detection in Germany. Its hearingis also acute, though its eyesight is comparatively weak, lacking colour vision and being unable to recognise a standing human 10–15 metres away.

Pigs are one of four known mammalian species which possess mutations in the nicotinic acetylcholine receptor that protect against snake venom. Mongooses, honey badgers, hedgehogs, and pigs all have modifications to the receptor pocket which prevents the snake venom α-neurotoxin from binding. These represent four separate, independent mutations.

Social Behavior and Life Cycle

Boars are typically social animals, living in female-dominated sounders consisting of barren sows and mothers with young led by an old matriarch. Male boars leave their sounder at the age of 8–15 months, while females either remain with their mothers or establish new territories nearby. Subadult males may live in loosely knit groups, while adult and elderly males tend to be solitary outside the breeding season **(Plate 115).**

Plate 115: Male Indian Boar (*S. s. cristatus*) Feeding on a Chital Carcass.

The breeding period in most areas lasts from November to January, though most mating only lasts a month and a half. Prior to mating, the males develop their subcutaneous armour, in preparation for confronting rivals. The testicles double in size and the glands secrete a foamy yellowish liquid. Once ready to reproduce, males travel long distances in search of a sounder of sows, eating little on the way. Once a sounder has been located, the male drives off all young animals and persistently chases the sows. At this point, the male fiercely fights potential rivals. A single male can mate with 5-10 sows. By the end of the rut, males are often badly mauled and have lost 20 per cent of their body weight. The gestation periodvaries according to the age of the expecting mother. For first time breeders, it lasts 114–130 days, while it lasts 133–140 days in older sows. Farrowing occurs between March and May, with litter sizes depending on the age and nutrition of the mother. The average litter consists of 4-6 piglets, with the maximum being 10-12. The piglets are whelped in a nest constructed from twigs, grasses and leaves. Should the mother die prematurely, the piglets are adopted by the other sows in the sounder.

Newborn piglets weigh around 600-1,000 grams, lacking underfur and bearing a single milk incisor and canine on each half of the jaw. There is intense competition between the piglets over the most milk-rich nipples, as the best fed young grow faster and have stronger constitutions. The piglets do not leave the lair for their first week of life. Should the mother be absent, the piglets lie closely pressed to each other. By two weeks of age, the piglets begin accompanying their mother on her journeys. Should danger be detected, the piglets take cover or stand immobile, relying on their camouflage to keep them hidden. The neonatal coat fades after three months, with adult colouration being attained at eight months. Although the lactation period lasts 2.5-3.5 months, the piglets begin displaying adult feeding behaviours at the age of 2–3 weeks. The permanent dentition is fully formed by 1–2 years. With the exception of the canines in males, the teeth stop growing during the middle of the fourth year. The canines in old males continue to grow throughout their lives, curving strongly as they age. Sows attain sexual maturity at the age of one year, with males attaining it a year later. However,estrus usually first occurs after two years in sows, while males begin participating in the rut after 4–5 years, as they are not permitted to mate by the older males. The maximum lifespan in the wild is 10–14 years, though few specimens survive past 4–5 years. Boars in captivity have lived for 20 years.

Habitat and Sheltering Behavior

The wild boar inhabits a diverse array of habitats from boreal taigas to deserts. In mountainous regions, it can even occupy alpine zones, occurring up to 1,900 metres in the Carpathians, 2,600 metres in the Caucasus and up to 3,600-4,000 metres in the mountains in Central Asia and Kazakhstan. In order to survive in a given area, wild boars require a habitat fulfilling three conditions: heavily brushed areas providing shelter from predators, water for drinking and bathing purposes and an absence of regular snowfall. The main habitats favoured by boars inEurope are deciduous and mixed forests, with the most favourable areas consisting of forest composed of oak and beech enclosing marshes andmeadows. In the Bia³owie¿a Forest, the animal's primary habitat consists of well developed, broad-leaved and mixed forests, along with marshy mixed forests, with coniferous forests and undergrowths being of secondary importance. Forests made up entirely of oak groves and beech are used only during the fruit-bearing season. This is in contrast to the Caucasian and Transcaucasian mountain areas, where boars will occupy such fruit-bearing forests year-round. In the mountainous areas of the Russian Far East, the species inhabits nutpine groves, hilly mixed forests where Mongolian oak and Korean pine are present, swampy mixed taiga and coastal oak forests. InTransbaikalia, boars are restricted to river valleys with nutpine and shrubs. Boars are regularly encountered in pistachiogroves in winter in some areas of Tajikistan and Turkmenia, while in spring they migrate to open deserts; boar have also colonised deserts in several areas they have been introduced to. On the islands of Komodo and Rinca, the boar mostly inhabits savanna or open monsoon forests, avoiding heavily forested areas unless pursued by humans. Wild boar are known to be competent swimmers, capable of covering long distances. In 2013, one boar was reported to have completed the seven mile swim from France to Alderney in the Channel Islands. Due to concerns about disease it was shot and incinerated.

Wild boar rest in shelters, which contain insulating material like spruce branches and dry hay. These resting places are occupied by whole families (though males lie separately), and are often located in the vicinity of streams, in swamp forests, in tall grass or shrub thickets. Boars never defecate in their shelters, and will cover themselves with soil and pine needles when irritated by insects.

Diet

The wild boar is a highly versatile omnivore, whose diversity in choice of food rivals that of humans. Its foods can be divided into four categories:

- ☆ Rhizomes, roots, tubers and bulbs, all of which are dug up throughout the year in the animal's whole range.

- ✫ Nuts, berries, and seeds, which are consumed when ripened and are dug up from the snow when abundant.
- ✫ Leaves, bark, twigs and shoots, along with garbage.
- ✫ Earthworms, insects, mollusks, fish, rodents, insectivores, bird eggs, lizards,snakes, frogs, and carrion. Most of these prey items are taken in warm periods.

A 50 kg boar needs around 4,000-4,500 calories of food per day, though this required amount increases during winter and pregnancy, with the majority of its diet consisting of food items dug from the ground like underground plant material and burrowing animals. Acorns and beechnuts are invariably its most important food items in temperate zones, as they are rich in thecarbohydrates necessary for the buildup of fat reserves needed to survive lean periods. In Western Europe, underground plant material favoured by boars includes bracken, willow herb, bulbs, meadow herb roots and bulbs, and the bulbs of cultivated crops. Such food is favoured in early spring and summer, but may also be eaten in autumn and winter during beechnut and acorn crop failures. Should regular wild foods become scarce; boars will eat tree bark andfungi, as well as visit cultivated potato and artichoke fields. Boar soil disturbance and foraging have been shown tofacilitate invasive plants. Boars of the *vittatus* subspecies in Ujung Kulon National Park in Java differ from most other populations by their primarily frugivorous diet, which consists of 50 different fruit species, especially figs, thus making them important seed dispersers. The wild boar can consume numerous genera of poisonous plants without ill effect, including*Aconitum*, *Anemone*, *Calla*, *Caltha*, *Ferula*, and *Pteridium*.

Boars may occasionally prey on small vertebrates like newborn deer fawns, leporids and galliform chicks. Boars inhabiting the Volga Delta and near some lakes and rivers of Kazakhstan have been recorded to feed extensively on fish like carp and Caspian roach. Boars in the former area will also feed on cormorant and heron chicks, bivalved molluscs, trapped muskrats and mice. There is at least one record of a boar killing and eating a bonnet macaque in southern India's Bandipur National Park, though this may have been a case of intraguild predation, brought on by interspecific competition for human handouts.

Diseases and Parasites

Wild boars are known to host at least 20 different parasitic worm species, with maximum infections occurring in summer. Young animals are vulnerable tohelminths like *Metastrongylus*, which are consumed by boars through earthworms, and cause death by parasitising the lungs. Wild boar also carries parasites known to infect humans, including *Gastrodiscoides*, *Trichinella spiralis*, *Taenia solium*, and*Balantidium coli*. Wild boar in southern regions are frequently infested with ticks(*Dermacentor*, *Rhipicephalus*, and *Hyalomma*) and hog lice. The species also suffers from blood-sucking flies, which it escapes by bathing frequently or hiding in dense shrubs.

Swine plague spreads very quickly in wild boar, with epizootics being recorded in Germany, Poland, Hungary, Belarus, the Caucasus, the Far East, Kazakhstan, and other regions. Foot-and-mouth disease can also take on epidemic proportions in boar populations. The species occasionally, but rarely contracts Pasteurellosis, hemorrhagic septicemia, tularemia andanthrax. Wild boar may on occasion contract swine erysipelas through rodents or hog lice and ticks

Crop and Garbage Raiding

Boars can be damaging to agriculture. Populations living on the outskirts of towns or farms can dig up potatoes and damage melons, watermelons and maize. They generally only encroach upon farms when natural food is scarce. In the Belovezh forest for example, 34-47 per cent of the local boar population will enter fields in years of moderate availability of natural foods. While the role of boars in damaging crops is often exaggerated, cases are known of boar depredations causing famines, as was the case in Hachinohe, Japan in 1749, where 3,000 people died of what became known as the 'wild boar famine'. Still within Japanese culture, the boar's status as vermin is expressed through its title as "king of pests" and the popular saying (addressed to young men in rural areas) "When you get married, choose a place with no wild boar." In Central Europe, farmers typically repel boars through distraction or fright, while in Kazakhstan it is usual to employ guard dogs in plantations. Although large boar populations can play an important role in limiting forest growth, they are also useful in keeping pest populations such as June bugs under control. The growth of urban areas and corresponding decline in natural boar habitats has led to some sounders entering human habitations in search of food. As in natural conditions, sounders in peri-urban areas are matriarchal, though males tend to be much less represented, and adults of both sexes can be up to 35 per cent heavier than their forest-dwelling counterparts. As of 2010, at least 44 cities in 15 countries have experienced problems of some kind relating to the presence of habituated wild boar

Attacks on Humans

Actual attacks on humans are rare, but can be serious, resulting in multiple penetrating injuries to the lower part of the body. They generally occur during the boars' rutting season from November –January, in agricultural areas bordering forests or on paths leading through forests. The animal typically attacks by charging and pointing its tusks towards the intended victim, with most injuries occurring on the thigh region. Once the initial attack is over, the boar steps back, takes position and attacks again if the victim is still moving, only ending once the victim is completely incapacitated.

Boar attacks on humans have been documented since the Stone Age, with the oldest depiction being a cave painting in Bhimbetaka, India. The Romans and Ancient Greeks wrote of these attacks (Odysseus was wounded by a boar), and several attacks are shown on the headstones of England's 12th century Severn Temple graveyard. A 2012 study compiling recorded attacks from 1825-2012 found accounts of 665 human victims of both wild boars and feral pigs, with the majority (19 per cent) of attacks in the animal's native range occurring in India. Most of the attacks occurred in rural areas during the winter months in non-hunting contexts, and were committed by solitary males

Management

1. Use of poison bait.
   (i) Spread the Orsenic oxide granules on flesh and keep it on the rout of boar.
   (ii) Fill the Orsenic oxide in holed apple and cover it and keep it on the rout of boar.
   (iii) Zinc phosphide coated with wax and covered with flour and further covered with apple or guava and fed to the boar.
2. Use of cages
3. Shooting
4. Use of electric fencing
5. Webbing of crop.
6. Use of explosive.
7. Destruction of boar sheds.
8. Watching.

Chapter 9

Nilgai/Blue Bull (*Boselaphus tragocamelus* Pallas, 1766)

Classification

Kingdom: Animalia

Phylum: Chordata

Class: Mammalia

Order: Artiodactyla

Family: Bovidae

Genus: *Boselaphus* Blainville, 1816

Species: *B. tragocamelus*

Binomial name: *Boselaphus tragocamelus* (Pallas, 1766)

The **nilgai** or **blue bull** (*Boselaphus tragocamelus*) is the largest Asian antelope and is endemic to the Indian subcontinent. The sole member of the genus *Boselaphus*, the species was described and given its binomial name by German zoologist Peter Simon Pallas in 1766. The nilgai stands 1–1.5 metres (3.3–4.9 ft) at the shoulder; males weigh 109–288 kilograms (240–635 lb), and the lighter females 100–213 kilograms (220–470 lb). A sturdy thin-legged antelope, the nilgai is characterised by a sloping back, a deep neck with a white patch on the throat, a short crest of hair along the neck terminating in a tuft, and white facial spots. A column of coarse hair, known as the "pendant", hangs from the dewlap ridge below the white patch. Sexual dimorphism is prominent – while females and juveniles are orange to tawny, adult males have a bluish greycoat. Only males possess horns, 15–24 centimetres (5.9–9.4 in) long.

The nilgai is diurnal (active mainly during the day). Three distinct kinds of groups are formed: one or two females with young calves, three to six adult and yearling females with calves, and all-male groups with two to 18 members. Typically tame, the nilgai may appear timid and cautious if harassed or alarmed; it flees up to 300 metres (980 ft)-or even 700 metres (2,300 ft), galloping away from the source of danger. Herbivores, nilgai prefer grasses and herbs, though woody plants are commonly eaten in the dry tropical forests of India. Females become sexually mature by two years, while males become sexually active only after four or five years. The time of the year when mating takes place varies geographically, but a peak breeding season lasting three to four months can be observed at most places. Gestation lasts eight to nine months, following which a single calf (sometimes twins or even triplets) is born. As typical of several bovid species, nilgai calves stay hidden for the first few weeks of their lives. The lifespan of the nilgai is around ten years.

Nilgai prefer areas with short bushes and scattered trees in scrub forests and grassy plains. They are common in agricultural lands, but hardly occur in dense forest. Major populations occur in the Terai lowlands in the foothills of theHimalayas (northern India), but the antelope is sparsely found in Nepal and Pakistan and is extinct in Bangladesh. Nilgai were first introduced to Texas in the 1920s and the 1930s. As of 2008, the feral population in Texas is nearly 37,000. The nilgai is categorised as Least Concern by the IUCN. The nilgai has been associated with Indian culture since the Vedic period (1500–500 BCE). Hindusrevere the nilgai as sacred and associate it with the cow, the mother animal inHinduism, through its name and loosely similar physical features. They were hunted in the Mughal era (16th to 19th centuries) and are depicted in numerous miniatures. Nilgai have been considered a pest in several north Indian states as they ravage crop fields and cause considerable damage. The nilgai has been declared as vermin in Bihar.

Etymology

The vernacular name "nilgai" comes from the fusion of the Hindi or Urdu *nil* ("blue") and *gai* (cow). The first recorded use of this word was in 1882. Alternative origins could be from the Sanskrit *gabhih* ("female bovid"), or Persian *gaw* ("cow"). The nilgai has been referred to by a variety of names: neelghae,[2] nilgau, nilgo, nylghau, nylghai and white-footed antelope. During Mughal emperor Aurangzeb's reign (1658–1707) over India, the nilgai was known by the name *nilghor* ("nil" for "blue" and "ghor" for "horse")

Description

The nilgai is the largest antelope in Asia. It stands 1–1.5 metres (3.3–4.9 ft) at the shoulder; the head-and-body length is typically between 1.7–2.1 metres (5.6–6.9 ft).[6] Males weigh 109–288 kilograms (240–635 lb); the maximum weight recorded is 308 kilograms (679 lb). Females are lighter, weighing 100–213 kilograms (220–470 lb). Sexual dimorphism is prominent; the males are larger than females and differ in colouration **(Plates 116-117).**

Plate 116: Nilgai – Male.

Plate 117: Female.

A sturdy thin-legged antelope, the nilgai is characterised by a sloping back, a deep neck with a white patch on the throat, a short mane of hair behind and along the back ending behind the shoulder, and around two white spots each on its face, ears, cheeks, lips and chin. The ears, tipped with black, are 15–18 centimetres (5.9–7.1 in) long. A column of coarse hair, known as the "**pendant**" and around 13 centimetres (5.1 in) long in males, can be observed along the dewlap ridge below the white throat patch. The tufted tail, up to 54 centimetres (21 in), has a few white spots and is tipped with black. The forelegs are generally longer, and the legs are often marked with white "socks".

While females and juveniles are orange to tawny, males are much darker – their coat is typically bluish grey. The ventralparts, the insides of the thighs and the tail are all white. A white stripe extends from the underbelly and broadens as it approaches therump, forming a patch lined with dark hair. Almost white, though not albino, individuals have been observed in the Sariska National Park(Rajasthan, India) while individuals with white patches have been recorded at zoos. The hairs, typically 23–28 centimetres (9.1–11.0 in) long, are fragile and brittle. Males have thicker skin on their head and neck that protect them in fights. The coat is not well-insulated with fat during winter, and consequently severe cold might be fatal for the nilgai.

Only males possess horns, though a few females may be horned as well. The horns are 15–24 centimetres (5.9–9.4 in) long but generally shorter than 30 centimetres (12 in). Smooth and straight, these may point backward or forward. The horns of the nilgai and the four-horned antelope lack the ringed appearance typical of those of other bovids.

The maximum recorded length of the skull is 376 millimetres (14.8 in). The milk teeth are totally lost and the permanent dentition completed by three years of age. The permanent teeth get degraded with age, showing prominent signs of wear at six years of age. The nilgai has sharp ears and eyes, though its sense of smell is not as acute.

Ecology and Behavior

The nilgai is **diurnal** (active mainly during the day). The daily routine of the antelope is feeding peaks at dawn, in the morning, in the afternoon and during the evening. Females and juveniles do not interact appreciably with males, except during the mating season. Groups are generally small, with ten or fewer individuals, though groups of 20 to 70 individuals can occur at times. The average herd size is of three individuals; herd membership varies with season. However, three distinct groupings are formed: one or two females with young calves, three to six adult and yearling females with calves, and male groups with two to 18 members.

Typically tame, the nilgai may appear timid and cautious if harassed or alarmed; instead of seeking cover like duikers it would flee up to 300 metres (980 ft)-or even 700 metres (2,300 ft) on galloping-away from the danger. Though generally quiet, nilgai have been reported to make short guttural grunts when alarmed, and females to make clicking noises when nursing young. Alarmed individuals, mainly juveniles below five months, give out a coughing roar (whose pitch is highest in case of the juveniles) that lasts half a second, but can be heard by herds less than 500 metres (1,600 ft) away and responded to similarly.

Fights take place in both sexes and involve pushing their necks against each other or ramming into one another using horns. Fights can be gory; despite the protective skin deep, lacerated wounds and even deaths might occur. Displaybehaviour focuses on the throat patch and the beard, and threatening opponents by pointing the horns toward them. A young male was observed making a submissive display in the Sariska Reserve by kneeling before an adult male, who stood erect. The nilgai mark their territories by forming dung piles as much as 50 centimetres (20 in) in radius. The defecation process is elaborate-the antelope stands with his legs about a metre apart, with the rump lowered and the tail held almost vertical; it stays in the same posture for at least ten seconds after relieving itself. The process is not as elaborate in the females as it is in the males.

In India, the nilgai shares its habitat with the four-horned antelope, chinkara, chital and blackbuck; its association with thegaur and the water buffalo is less common. In the Ranthambore National Park (Rajasthan, India) the nilgai and the chinkara collectively prefer the area rich in *Acacia* and *Butea* species, while the sambar deer and the chital preferred the forests of *Anogeissus* and *Grewia* species. In India, the Indian tiger and the lion may prey on the nilgai but they are not significant predators of this antelope. Leopards prey on the nilgai, though they prefer smaller prey. Dholes generally attack juveniles. Other predators include wolves and striped hyenas

Diet

Herbivores, the nilgai prefer grasses and herbs; woody plants are commonly eaten in the dry tropical forests of India. Studies suggest they may be browsers[l]or mixed feedersin India, whereas they are primarily grazers in Texas. The nilgai can tolerate interference by livestock and degradation of vegetation in its habitat better than deer, possibly because they can reach high branches and do not depend on surface vegetation.[40] The sambar deer and nilgai in Nepal have similar dietary preferences. Diets generally suffice in protein and fats. The protein content of the nilgai's should be at least seven percent. The nilgai can survive for long periods without water and does not drink regularly even in summer. However, a nilgai died in Dwarka (India) allegedly due to the heat wave and acute shortage of water.

Preference for herbs and grasses; grasses became more important in the rainy season, while during winter and summer it fed additionally on flowers (*Butea monosperma*), foliage (*Anogeissus pendula, Capparis sepiaria, Grewia flavescens* and *Zizyphus mauritiana*), pods (*Acacia nilotica, A. catechu* and A. leucophloea) and fruits (*Zizyphus mauritiana*). Preferred grass species include *Cenchrus* species,*Cynodon dactylon, Desmostachya bipinnata, Scirpus tuberosus* and *Vetiveria zizanoides*. Woody plants eaten include*Acacia nilotica, A. senegal, A. leucophloea, Clerodendrum phlomidis, Crotalaria burhia, Indigofera oblongifolia, Morus alba*and *Zizyphus nummularia*; herbs favoured are *Cocculus hirsutus, Euphorbia hirta* and *Sida rhombifolia*. Seeds of*Paspalum distichum* occurred in the dung of nilgai most of the year; *Acacia nilotica* and *Prosopis juliflora* seeds were discovered in the dry season and those of *Echinochloa crusgalli* during the monsoon

Reproduction

Observations of females in southern Texas revealed that ovaries are developed by two years of age and the first birth takes place typically a year later, though in a few cases females as young as one-and-a-half years may mate successfully. Females can breed again around a year after parturition. Males in the same location were found to have active testes by the age of three years, that maturedconsiderably by the next year. Males become sexually active at four or five years. Mating may occur throughout the year, with peaks of three to four months. The time of the year when these peaks occur varies geographically. In Texas, a peak is apparent from December to March.[32] In Bharatpur National Park (Rajasthan, India) the breeding season is from October to February, peaking in November and December. The Sariska reserve witnesses a similar peak in December and January

In the mating season, rutting males move about in search of females in oestrus. Males become aggressive and fight among themselves for dominance. These fights are characterised by displays of the enlarged chest, the throat patch and the beard while holding the head upright; and threatening the opponent by running with the horns pointed toward him and circling him. The victorious bull would protect the vicinity of the targeted female from other males. The courtship typically lasts for 45 minutes. The male, stiff and composed, approaches the receptive female, who keeps her head low to the ground and may slowly walk forward. The male licks her genitalia, upon which the female holds her tail to a side and the male gives out a flehmen response. Finally, the male pushes his chest against her rump, and mounts her.

Gestation lasts eight to nine months, following which a single calf or twins (even triplets at times) are born. In a 2004 study in the Sariska reserve, twins accounted for as high as 80 percent of the total calf population. Births peak from June to October in the Bharatpur National Park, and from April to August in southern Texas. Calves are precocial; they are able to stand within 40 minutes of birth, and forage by the fourth week. Pregnant females isolate themselves before giving birth. As typical of several bovid species, nilgai calves are kept in hiding for the first few weeks of their lives. This period of concealment can last as long as a month in Texas. Calves, mainly males, bicker playfully by neck-fighting. Young males would leave their mothers at ten months to join bachelor groups. The lifespan of the nilgai is typically ten years in Texas.

Habbit and Distribution

Nilgai prefer areas with short bushes and scattered trees in scrub forests and grassy plains. They are common in agricultural lands, but hardly occur in dense woods. In southern Texas, it roams in the prairies, scrub forests and oak forests. It is a generalist animal-it can adapt to a variety of habitats. Though sedentary and less dependent on water, nilgai may desert their territories if all water sources in and around it dry up.

This **antelope** is endemic to the Indian subcontinent: major populations occur in India, Nepal and Pakistan, whereas it is extinct in Bangladesh. Significant numbers occur in the Terai lowlands in the foothills of the Himalayas; the antelope is abundant across northern India. The Indian population was estimated at one million in 2001. The nilgai were first introduced to Texas in the 1920s and the 1930s in a 6,000 acres (2,400 ha) large ranch near the Norias Division of theKing Ranch, one of the largest ranches in the world. The feral population saw a spurt toward the latter part of the 1940s, and gradually spread out to adjoining ranches.

Population densities show great geographical variation across India. Density can be as low as 0.23 to 0.34 individuals per km^2 in the Indravati National Park(Chhatisgarh) and 0.4 individuals per km^2 in the Pench Tiger Reserve (Madhya Pradesh) or as high as 6.60 to 11.36 individuals per km^2 and Ranthambhore National Park and 7 individuals per km^2 in Keoladeo National Park (both in Rajasthan). Seasonal variations were noted in the Bardiya National Park (Nepal) in a 1980 study; the density 3.2 individuals per km^2 during the dry season and 5 per km^2 in April (the start of the dry season). In southern Texas, densities were found to be nearly 3–5 individuals per km^2 in 1976.

Historic notes mention nilgai in southern India, but these may have been feral:

It is believed that the Coimbatore and Salem collectorates are almost the only places in Southern India, in which nil-gai are to be found. It is difficult to account for the animals being thus so widely divided from their usual haunts unless as has been generally supposed, these Southern specimens are the progeny of a semi-domesticated herd, which, at some by-gone period, had escaped from the preserve of a native potentate

Conservation

The nilgai is protected under Schedule III of the Wildlife Protection Act of 1972. Major protected areas for the nilgai across India include: Gir National Park (Gujarat); Bandhavgarh National Park, Bori Wildlife Sanctuary, Kanha National Park, Pachmarhi Biosphere Reserve, Panna Tiger Reserve, Pench Tiger Reserve,Sanjay National Park, Satpura National Park (Madhya PRadesh); Tadoba Andhari Reserve

(Maharashtra); Kumbhalgarh Wildlife Sanctuary, Ranthambore National Park and Sariska Tiger Reserve (Rajasthan)

Damage

The populations of nilgai in India are so large that farmers in the states of Bihar, Chhattisgarh, Haryana, Madhya Pradesh, Maharashtra, Rajasthan and Uttar Pradesh have pleaded to the government to cull them. Nilgai herds raid and trample crop fields across these states, often causing food shortages. Farmers use live electric wires to guard their farms, which kills other animals as well. Farmers in Neemuch (Madhya Pradesh) went on a hunger strikein 2015 demanding compensation for the damage caused by nilgai. Although black buck cause a similar problem, the damage caused by them is significantly lower as they merely break off young shoots. A 1990 study suggested culling, building enclosures for the antelopes and fencing off agricultural areas as remedies for the menace.

The governments of Bihar, Maharashtra and Uttarakhandhave urged the Government of India to declare the nilgai as a vermin; the proposal has been implemented in Bihar, where nilgai can now be hunted to minimise the damages incurred by locals. The Uttar Pradesh government has given farmers and firearm licence holders the right to cull the animals, However, animal rights activists in Maharashtra were unhappy with the decision. The state governments have attempted other initiatives to curb the nilgai menace: in November 2015, the Rajasthan government came up with a proposal to allow shooting nilgai with non-lethal darts to inhibit fertilisation in their bodies, so as to regulate their increasing populations. As the name "nilgai" appeals to the religious sentiments of Hindus, the Government of Madhya Pradeshhas sought to officially rename it as *rojad* (Hindi for "forest antelope") and the Government of Haryana to rename it as*roze* in a bid to make their culling acceptable.

A 1994 study drew attention to the ecological value provided by the nilgai in ravines lining the Yamuna River. In summer, the faeces of the antelope contained nearly 1.6 percent nitrogen that could enhance the quality of the soil up to a depth of 30 centimetres (12 in). Seeds in the droppings could easily germinate and assist in afforestation.

Chapter 10

Crop Raiding Elephants (*Elephas maximus indicus* Cuvier, 1798)

Classification

Kingdom: Animalia

Phylum: Chordata

Class: Mammalia

Order: Proboscidea

Family: Elephantidae

Genus: *Elephas*

Species: *Elephas maximus*

Binomial name: *Elephas maximus indicus* (Cuvier), 1798

The **Indian elephant** (*Elephas maximus indicus*) is one of three recognizedsubspecies of the Asian elephant and native to mainland Asia. Since 1986, *Elephas maximus* has been listed as *Endangered* by IUCN as the population has declined by at least 50 per cent over the last 1 to -1 years or three generations. Asian elephants are threatened by habitat loss, degradation and fragmentation.

Characteristics

In general, Asian elephants are smaller than African elephants and have the highest body point on the head. The tip of their trunk has one finger-like process.

Plate 118: Indian Elephant Tusked Male.

Plate 119: Female.

Their back is convex or level. Indian elephants reach a shoulder height of between 2 and 3.5 m (6.6 and 11.5 ft), weigh between 2,000 and 5,000 kg (4,400 and 11,000 lb), and have 19 pairs of ribs. Their skin color is lighter than of*maximus* with smaller patches of depigmentation, but darker than of *sumatranus*. Females are usually smaller than males, and have short or no tusks.

The largest Indian elephant was 3.43 metres (11.3 ft) high at the shoulder. In 1985, two large elephant bulls were spotted for the first time in Bardia National Park, and named *Raja Gaj* and *Kanchha*. They roamed the park area together and made occasional visits to the females. *Raja Gaj* stood 11.3 ft (3.4 m) tall at the shoulder and had a massive body weight. His appearance has been compared to that of a mammoth due to his high bi-domed shaped head. His forehead and domes were more prominent than in other Asian bull elephants **(Plates 118-119)**.

Indian elephants have smaller ears, but relatively broader skulls and larger trunks than African elephants. Toes are large and broad. Unlike their African cousins, their abdomen is proportionate with their body weight but the African elephant has a large abdomen as compared to the skulls

Distribution and Habitat

Indian elephants are native to mainland Asia: India, Nepal, Bangladesh, Bhutan,Myanmar, Thailand, Malay Peninsular, Laos, China, Cambodia, and Vietnam. They are regionally extinct from Pakistan.[1] They inhabit grasslands, drydeciduous, moist deciduous, evergreen and semi-evergreen forests. In the early 1990s, their estimated population size was

- 27,785-31,368 in India, where populations are restricted to four general areas:
- in the Northwest – at the foot of the Himalayas in Uttarakhand and Uttar Pradesh, ranging from Katarniaghat Wildlife Sanctuary to the Yamuna River;
- in the Northeast – from the eastern border of Nepal in northern West Bengal through western Assam along the Himalaya foothills as far as theMishmi Hills, extending into eastern Arunachal Pradesh, the plains of upper

Assam, and the foothills of Nagaland, to the Garo Hills of Meghalayathrough the Khasi Hills, to parts of the lower Brahmaputra plains and Karbi Plateau; isolated herds occur in Tripura, Mizoram, Manipur, and in theBarak Valley districts of Assam:

- in the central part – in Odisha, Jharkhand, and in the southern part of West Bengal, with some animals wandering into Chhattisgarh;
- in the South – eight populations are fragmented from each other in northern Karnataka, in the crestline of Karnataka–Western Ghats, inBhadra–Malnad, in Brahmagiri–Nilgiris–Eastern Ghats, in Nilambur–Silent Valley–Coimbatore, in Anamalai–Parambikulam, in Periyar–Srivilliputhur, and one in Agasthyamalai;
- 100–125 in Nepal, where their range is restricted to a few protected areas in the Terai along the border with India. In 2002, estimates ranged from 106 to 172 resident and migratory elephants, with most of them in Bardia National Park;
- 150–250 in Bangladesh, where only isolated populations survive in theChittagong Hills;
- 250–500 in Bhutan, where their range is limited to protected areas in the south along the border with India;
- 4,000–5,000 in Myanmar, where populations are highly fragmented, and occur in the northern and western hill ranges, in Pegu Yoma of central Myanmar,Tenasserim and Shan State;
- 2,500–3,200 in Thailand, mainly in the mountains along the border with Myanmar, with smaller fragmented populations occurring in the peninsula in the south;
- 2,100–3,100 in Malaysia;
- 500–1,000 Laos, where they remain widely but patchily distributed in forested areas, both in the highlands and lowlands;
- 200–250 in China, where they survive only in the prefectures ofXishuangbanna, Simao, and Lincang of southern Yunnan;
- 250–600 in Cambodia, where they primarily inhabit the mountains of the south-west and in Mondulkiri and Ratanakiri Provinces;
- 70–150 in the southern parts of Vietnam.

Ecology and Behavior

Elephants are classified as megaherbivores and consume up to 150 kg (330 lb) of plant matter per day. They are generalist feeders, and both grazers andbrowsers. In a study area of 1,130 km^2 (440 sq mi) in southern India, elephants were recorded to feed on 112 different plant species, most commonly of the orderMalvales, and the legume, palm, sedge and true grass families. They graze on thetall grasses, but the portion consumed varies with season. When the new flush appears in April, they remove the tender blades in small clumps. Later, when grasses are higher than 0.5 m (1.6 ft), they uproot entire clumps, dust them skilfully and consume the fresh leave tops, but discard the roots. When grasses are mature in autumn, they clean and consume the succulent basal portions with the roots, and discard the fibrous blades. From the bamboos, they eat seedlings, culms and lateral shoots. During the dry season from January to April, they mainly browse on both leaves and twigs preferring the fresh foliage, and consume thorn bearing shoots of acacia species without any obvious discomfort. They feed on the bark of white thorn and other flowering plants, and consume the fruits of wood apple, tamarind, kumbhi and date palm.

In Nepal's Bardia National Park, elephants consume large amounts of the floodplain grass, particularly during themonsoon season. They browse more in the dry season with bark constituting a major part of their diet in the cool part of that season. During a study in a tropical moist mixed deciduous forested area of 160 km^2 (62 sq mi) in Assam, elephants were observed to feed on about 20 species of grasses, plants and trees. Grasses such as *Imperata cylindrica* and *Leersia hexandra* constituted by far the most predominant component of their diet.

The movement and habitat utilisation patterns of an elephant population were studied in southern India during 1981–83 within a 1,130 km^2 (440 sq mi) study area. The vegetation types of this area encompasses dry thorn forest at 250 to 400 m (820 to 1,310 ft), deciduous forest at 400 to 1,400 m (1,300 to 4,600 ft), stunted evergreen forest and grassland at 1,400 to 1,800 m (4,600 to 5,900 ft). Five different elephant clans, each consisting of between 50 and 200 individuals had home ranges of between 105 km^2 (41 sq mi) and 320 km^2 (120 sq mi), which overlapped. They preferred habitat where water was available and food plants were palatable. During the dry months of January to April, they congregated at high densities of up to five individuals per km^2 in river valleys where browse plants had a much higher protein content than the coarse tall grasses on hill slopes. With the onset of rains in May, they dispersed over a wider area at lower densities, largely into the tall grass forests, to feed on the fresh grasses, which then had a high protein value. During the second wet season from September to December, when the tall grasses became fibrous, they moved into lower elevation short grass open forests. The normal movement pattern could be upset during years of adverse environmental conditions. However, the movement pattern of elephants in this region has not basically changed for over a century, as inferred from descriptions recorded during the 19th century.

In the Nilgiri Biosphere Reserve three elephant clans had overall home ranges of 562 km^2 (217 sq mi), 670 km^2(260 sq mi) and 799 km^2 (308 sq mi) in the beginning of the 1990s. During three years of survey, their annual home ranges overlapped to a large extent with only minor shifts in the home ranges between years.

References

Mites

Andrews RM, McCarthy J, Carapetis JR and Currie BJ (2009). "Skin disorders, including pyoderma, scabies, and tinea infections". Pediatr. Clin. North Am. **56** (6): 1421– 40. doi:10.1016/j.pcl.2009.09.002. PMID 19962029.

Arroyo, J.; Keith, A.M.; Schmidt, O. and Bolger, T. (2013). "Mite abundance and richness in an Irish survey of soil biodiversith with comments on some newly recorded species". *Ir Nat. J* **33**: 19–27.

Ashutosh Sharma and Pratap Kumar Pati (2012). "First record of the carmine spider mite, Tetranychus urticae, infesting *Withania somnifera* in India" (PDF). *Journal of Insect Science* **12**: *50.* doi:10.1673/031.012.5001.

B. A. Croft, P. D. Pratt, G. Koskela and D. Kaufman (1998). "Predation, reproduction, and impact of phytoseiid mites (Acari: Phytoseiidae) on cyclamen mite (Acari: Tarsonemidae) on strawberry". *Journal of Economic Entomology*, 91 (6): 1307 1314.

D. A. Raworth, D. R. Gillespie, M. Roy and H. M. A. Thistlewood (2002). "Tetranychus urticae Koch, twospotted spider mite (Acari: Tetranychidae)". In Peter G. Mason and John Theodore Huber. *Biological Control Programmes in Canada,* 1981–2000.CAB International. pp. 259–265. ISBN 978-0-85199-527-4.

D. E. Walter and H. C. Proctor (1999). Mites: Ecology, Evolution and Behaviour. University of NSW Press, Sydney and CABI, Wallingford. ISBN 0-86840-529-9. Of Knots and Worms Not: Gordialycus.

David Evans Walter, Gerald Krantz and Evert Lindquist (1996). "Acari. The Mites". Tree of Life Web Project. Retrieved June 21, 2010.

Diaz, J. H. (2010). "Endemic mite-transmitted dermatoses and infectious diseases in the South". The Journal of the Louisiana State Medical Society: official organ of the Louisiana State Medical Society **162** (3): 140–5, 147–9. PMID 20666166.

G. Sterk, G. E. Bal, W. Goossens and D. Bylemans (1997). "Semi-field and field experience in the control of the strawberry mite, *Tarsonemus pallidus* (Banks) (Acarina: Tarsonemidae)". Parasitica 53 (1): 25 –33.

Gerald W. Krantz and D. E. Walter, ed. (2009). A Manual of Acarology (3rd ed.). Texas Tech University Press. ISBN 978-0-89672-620-8.

Graham Bell (1982). "Parthenogenesis and vegetative reproduction in multicellular animals". *The Masterpiece of Nature: the Evolution and Genetics of Sexuality*. Croom Helm applied biology series. Cambridge University Press. pp. 160–331. ISBN 978-0-85664-753-6.

Günther Schmidt (1993). Giftige und gefährliche Spinnentiere (in German). Westarp Wissenschaften. pp. 58ff. ISBN 3-89432-405-8.

H. R. Bolland, Jean Gutierrez and Carlos H. W. Flechtmann (1997). "Introduction". World Catalogue of the Spider Mite Family (Acari: Tetranychidae). Brill Publishers. pp. 1– 3. ISBN 978-90-04-11087-8.

https://en.wikipedia.org/wiki/Acari - cite_ref-9Barker, S.C. and Murrell, A. (2004). "Systematics and evolution of ticks with a list of valid genus and species names". *Parasitology*, **129**: S15–S36.doi:10.1017/S0031182004005207.

https://en.wikipedia.org/wiki/Acari - cite_ref-Garw_2-0Garwood, Russell J.; Dunlop and Jason A. (2014). "Three-dimensional reconstruction and the phylogeny of extinct chelicerate orders". PeerJ **2**: e641.doi:10.7717/peerj.641. Retrieved June 15, 2015.

https://en.wikipedia.org/wiki/Spider_mite - cite_ref-6 *H. R. Bolland, Jean Gutierrez and Carlos H. W. Flechtmann (1997). "Key to the genera of the world"*. World Catalogue of the Spider Mite Family (Acari: Tetranychidae). Brill Publishers. *pp. 5-11.* ISBN 978-90-04-11087-8.

https://en.wikipedia.org/wiki/Tarsonemidae - cite_ref-6 *P. G. Weintraub and E. Palevsky (2003). "Distribution and diel movement of the predatory mite, Neoseiulus cucumeris, on greenhouse sweet pepper – preliminary study".IOBC/WPRS Bulletin* ***26****: 89–94.*

https://en.wikipedia.org/wiki/Tetranychus_urticae - cite_ref-4*Boran Altincicek, Jennifer L. Kovacs and Nicole M. Gerardo (2011).* "Horizontally transferred fungal carotenoid genes in the two-spotted spider mite Tetranychus urticae". *Biology Letters* **8** (2): 253–257. doi:10.1098/rsbl.2011.0704. PMC 3297373.PMID 21920958.

L. R. Jeppson, Hartford H. Keifer and Edward William Baker (1975). "The Tarsonemidae Kramer". *Mites injurious to economic plants*. University of California Press. pp. 285– 306. ISBN 978-0-520-02381-9.

Michael Heethoff and Lars Koerner (2007). "Small but powerful – the oribatid mite Archegozetes longisetosus Aoki (Acari, Oribatida) produces disproportionate high forces". *Journal of Experimental Biology*, **210** (17): 3036 3042. doi:10.1242/jeb. 008276. PMID 17704078.

Nugroho and Y. Ibrahim (2004). "Laboratory bioassay of some entomopathogenic fungi against broad mite". *Journal of Agricultural Biology*, 6 (2): 223-225.

Nyle C. Brady and Weil, Ray R. (2009). Elements of the Nature and Properties of Soils (3rd Edition). Prentice Hall. ISBN 9780135014332.

R. B. Halliday, B. M. OConnor and A. S. Baker (2000). "Global Diversity of Mites". In Peter H.

Raven and Tania Williams. Nature and human society: the quest for a sustainable world: proceedings of the 1997 Forum on Biodiversity. *National Academies*. pp. 192–212.

Thomas R. Fasulo and H. A. Denmark (2009). "Two spotted spider mite". Featured Creatures. University of Florida/Institute of Food and Agricultural Sciences. Retrieved May 20, 2011.

Tien NS, Massourakis G, Sabelis MW and Egas M. (2011). "Mate choice promotes inbreeding avoidance in the two-spotted spider mite". *Exp. Appl. Acarol.* **54** (2): 119– 24. doi:10.1007/s10493-011-9431-y. PMC 3084432. PMID 21400191.

Uesugi, R.; Goka, K.and Osakabe, M. H. (2002). "Genetic Basis of Resistances to Chlorfenapyr and Etoxazole in the Two-Spotted Spider Mite (Acari: Tetranychidae)". *Journal of Economic Entomology*,**95** *(6): 1267–1274.* doi:10.1603/0022-0493-95.6.1267. ISSN 0022- 0493. PMID 12539841.

Villa, Scott M; Bohec, Celine Le (October 2013). "Diversity of Feather Mites (Acari: Astigmata) on Darwin's Finches".*The Journal of Parasitology*. doi:10.1645/12-112.1.

Yutaka Saito (2009). "Plant mites". Plant Mites and Sociality: Diversity and Evolution. Springer. pp. 5–38.doi:10.1007/978-4-431-99456-5_2. ISBN 978-4-431-99455-8.

Root-knot Nematode

Ali S.S. (1986). Root-knot nematode problem in cardamom and its management. Proceedings of the Second Group Discussions on the Nematological Problems of Plantation Crops, April 24- 25, 1986. Central Coffee Research Station, Balehonnur, Karnataka, India, pp. 10-12.

Amin A.W. and Budai C.S. (1994). Some weed host plants of the root-knot nematode Meloidogyne species in south-eastern Hungary. *Pakistan Journal of Nematology*, 12(1):59-65.

Anamika, Sobita Simon (2010). New report on occurrence of root-knot disease in Beta vulgaris. Current Nematology, 21(1/2):71-73.

Antonio H. (1988). Avaliacao das perdas causadas por Meloidogyne incognita raca 4 no cultivar BR-4 de soja. Nematologia Brasileira, 12:29-34.

Araki M and Nakasono K.A. (1987). Possible record of *Meloidogyne incognita* race 3 from Japan. *Japanese Journal of Nematology*, 17(12):59-60.

Artyunov A.V. (1985). Allium pskemense - a new plant host of Meloidogyne. *Izvestiya Akademii Nauk Turkmenskoi SSR Biologicheskie Nauk,* No.7, 64-65.

Babatola J.D. (1980). Reactions of some rice cultivars to the root-knot nematode, *Meloidogyne incognita. Nematropica,* 10(1):5-9.

Babatola J.O. and Omotade M.A. (1991). Chemical control of the nematode pests of cowpea, *Vigna unguiculata* (L.) Walp. *Crop Protection,* 10(2):121-124;

Bahmani J., Khozeini F., Barooti S., Rezaee S., and Ghaderi R, (2013). Plant-parasitic nematodes associated with walnut in the Sanandej region of west Iran. *Journal of Plant Protection Research,* 53(4):404-408. http://www. degruyter.com/view/j/jppr.2013.53.issue-4/jppr- 2013- 0060/jppr-2013-0060.xml?format=INT.

Bala G (1984). Occurrence of plant-parasitic nematodes associated with crops of agricultural importance in Trinidad. *Nematropica,* 14(1):37-45.

Barker K.R., Todd F.A., Shane W.W., and Nelson L.A. (1981). Interrelationships of Meloidogyne species with flue-cured tobacco. *Journal of Nematology,* 13(1):67-79.

Batyr A.K. and Kozhokaru G.I. (1985). Distribution and species composition of gall nematodes in commercially grown flowers in Moldavia. Ekologiya i prakticheskoe znachenie zoo- i fitoparaziticheskikh organizmov. *Kishinev,* USSR, Shtiintsa., 3-10,

Bhardwaj L.N. and Hogger C.H. (1984). Root-knot nematodes of Chitwan District of Nepal. *Nematologia Mediterranea,* 12(1):155-158

Bhat O.K. and Kaul V. (1994). Influence of soil temperature on the population behaviour of root- knot nematode, Meloidogyne incognita in a tomato field. *Annals of Plant Protection Sciences,* 2(1):28-29; 6 ref.

Bhattacharya C., Dasgupta M.K. and Mukherjee B. (2012). Biodiversity of plant parasitic nematodes of cashew plantations in Tripura, India. *Nematologia Mediterranea,* 40(2):163- 168. http://www.edizioniets.com

Bhatti DS, Jain RK, 1977. Estimation of loss in okra, tomato and brinjal yield due to Meloidogyne incognita. *Indian Journal of Nematology,* 7(1):37-41

Bora A. (1970). Studies on plant-parasitic nematodes in the Black Sea region and their distribution and possibilities for chemical control. *Bitki Koruma Bulteni,* 10(1):53-71

Bridge J. (1988). Plant-parasitic nematode problems in the Pacific Islands. *Journal of Nematology,* 20(2):173-183.

Bridge J. (1992). Nematodes. In: Hillocks RJ, ed. Cotton Diseases, Wallingford, UK: *CAB Internationa*l, 331-353.

Bridge J., Otim-Nape W., Namaganda J. (1991). The root-knot nematode, *Meloidogyne incognita,* causing damage to cassava in Uganda. *Afro-Asian Journal of Nematology,* 1(1):116-117

Campos V.P., Sivapalan P. and Gnanapragasam N.C. (1990). Nematode parasites of coffee, cocoa and tea. In: Luc M, Sikora RA, Bridge J, eds. Plant Parasitic Nematodes in Subtropical and Tropical Agriculture. Wallingford, UK: *CAB International*, 387-430.

Chen C.M., Chen P. and Yao G. (1991). Study on root-knot nematodes on tobacco and their control. *Sichuan Plant Protection*, No. 1/2, 75-78.

Chen C.M., Li H.Y. and Lii D.Y. (1986). The study on root-knot nematodes of common turmeric (*Curcuma domestica* Valet). *Herald of Agricultural Sciences*, 1:16-22.

Chunram C. (1972). A list of plant parasitic nematodes in Thailand. Plant Protection Service Technical Bulletin, Ministry of Agriculture, Bangkok, Thailand, No. 1:44 pp.

Colombo A., Cataldi S., Marano G. and Genna G.(2008). Soil solarization and phenamiphos to control the root-knot nematode Meloidogyne incognita in Sicily. Redia [IX Congress of the Italian Society of Nematology (SIN), Florence, Italy, 22-24 November 2007.], 91:97-102.

Cuadra R., Aguilera C., PTrez J.A. and Guerra L.C. (1988). Host plants of Meloidogyne incognita in Granma province, Cuba. *Ciencias de la Agricultura*, No. 34-35:18-23; 12 ref.

Dabaj K.H. and Jenser. (1987). List of plants infected by root-knot nematodes in Libya. *International Nematology Network Newsletter*, 4(3):28-33

De Vito M., Greco N. and Carella A. (1986). Effect of Meloidogyne incognita and Importance of the Inoculum on the Yield of Eggplant. *Journal of Nematology*, 18:487-490.

Diez and Dusenbery. (1989). Repellent of root-knot nematodes from exudate of host roots. *J. Chem. Ecol.* 15:2445-2455.

Diomandé M. (1981). Effets direct et combin, des engrais et de Meloidogyne incognita sur le riz pluvial. *Revue de Nématologie*, 4:71-74.

Diomande M. (1984). Response of upland rice cultivars to Meloidogyne species. *Revue de Nematologie*, 7(1):57-63

Doucet M.E. and Pinochet J. (1992). Occurrence of Meloidogyne spp. in Argentina. *Journal of Nematology*, 24(4, Supplement):765-770; 60 ref.

Dusenbery. (1987). Theoretical range over which bacteria and nematodes locate plant roots using carbon dioxide. *J. Chem. Ecol.* 13: 1617-1624.

Eisenback, J. D. and Triantaphyllou, H. H. (1991). Root-knot Nematodes: Meloidogyne species and races. In: Manual of Agricultural Nematology, W. R. Nickle. (Ed). Marcel Dekker, New York. pp 281 – 286.

Eissa, M.F.M. (1982). Species and race identification of the root-knot nematodes, Meloidogyne species from four regions of the kingdom of Saudi Arabia. *Agricultural Research Bulletin, Faculty of Agriculture, Al-Azhar University*, Cairo, No. 11:11 pp.

Eissa, M.F.M. and Hyder S.Z. 1981. New host records for the root-knot nematodes, *Meloidogyne javanica* and *M. incognita* race II from *Saudi Arabia. Research Bulletin, Faculty of Agriculture, Zagazig University, Egypt*, No. 407:8 pp.

El-Sherbiny A.A. (2011). Phytoparasitic nematodes associated with ornamental shrubs, trees and palms in Saudi Arabia, including new host records. *Pakistan Journal of Nematology*, 29(2):147-164.

Errico G., Crescenzi A. and Landi S. (2014). First report of the southern root-knot nematode *Meloidogyne incognita* on the invasive weed Araujia sericifera in Italy. Plant Disease, 98(11):1593-1594. http://apsjournals.apsnet.org/loi/pdis.

Esfahani M.N. and Ahmadi A. (2010). Field observations on the reaction of medicinal plants to root-knot nematodes in Isfahan, Iran. International Journal of Nematology, 20(1):107-112.

Fademi O.A. (1984). Influence of rate and time of carbofuran application to control root-knot nematodes in upland rice. *International Rice Research Newsletter*, 9:22-23.

Fademi O.A. (1986). Reaction of some selected rice cultivars to a root-knot nematode, *Meloidogyne incognita* (Kofoid and White) Chitwood. *Pakistan Journal of Nematology*, 4:15-18.

Fademi O.A. (1987). Resistance of some rice varieties to the root-knot nematode (RKN) *Meloidogyne incognita. International Rice Research Newsletter*, 12:11.

Fam Tkhan' Bin' (1990). Gall nematodes of vegetables and potatoes in Da Lat (Tei Nguen Plateau, Vietnam) and description of *Meloidogyne cynariensis*, a parasite of artichokes. *Zoologicheskii^breve~ Zhurnal*, 69(4):128-131.

Ferris H. and Van Gundy S.D. (1979). Meloidogyne ecology and host interrelationships. Root-knot nematodes (Meloidogyne species); systematics, biology and control. Academic Press Inc. London and New York, USA UK, 205-230.

Garcia O. and Perez M, 1987. PTrdidas en cosecha producidas por Meloidogyne incognita en tabaco de la provincia de Pinar del Rfo. *Ciencia y Tenica en la Agricultura. Protection de Plantas*, 10(3):85-98.

Gharabadiyan F., Jamali S., Yazdi A.A., Hadizadeh M.H. and Eskandari A.(2012). Weed hosts of root-knot nematodes in tomato fields. *Journal of Plant Protection Research*, 52(2):230-234. http://versita. metapress. com/link.asp?target= contribution and id=428087616X360N23.

Hague N.G.M., Gowen S.R. (1987). Chemical Control of Nematodes. In: Brown RH, Kerry BR, eds. Principles and Practice of Nematode Control in Crops. London, UK: Academic Press, 131 - 178.

https://en.wikipedia.org/wiki/Root-knot_nematode - cite_ref-10*Charles, Lauren., Carbone, Ignazio., Davies, Keith G., Bird, David., Burke, Mark., Kerry, Brian R., and Opperman, Charles H. (2005)* "Phylogenetic Analysis of Pasteuria penetrans by Use of Multiple Genetic Loci". *Journal of Bacteriology* ***187***.doi:10.1128/JB.187.16.5700- 5708.2005. Retrieved 2014-09-13.

https://en.wikipedia.org/wiki/Root-knot_nematode - cite_ref-11*Dama, L.B., Poul, B.N., Jadhav, B.V. and Hafeez, M.D. (1999). "Effect of Herbal "Juglone" on Development* *of the plant parasitic nematode (Meloidogyne Spp.) on Arachis hypogaea". Journal of Ecotoxicology and Environmental Monitoring,* ***9****: 73–76.*

https://en.wikipedia.org/wiki/Root-knot_nematode - cite_ref-12 Sijmons, P. C., Grundler, F. M. W., vonMende, N., Burrows, P. R. and Wyss, U. (1991). Arabidopsis thalliana as a new model host for plant-parasitic nematodes. *The Plant Journal,* 1, 245 - 254.

https://en.wikipedia.org/wiki/Root-knot_nematode - cite_ref-13Wyss, U., Grundler, F.M.W. and Munch, A. (1992). The parasitic behaviour of second stage juveniles of Meloidogyne incognita in roots of Arabidopsis thaliana. Nematologica, 38, 98 - 111.https://en.wikipedia.org/wiki/Root-knot_nematode - cite_ref-24

https://en.wikipedia.org/wiki/Root-knot_nematode - cite_ref-14

https://en.wikipedia.org/wiki/Root-knot_nematode - cite_ref-15

https://en.wikipedia.org/wiki/Root-knot_nematode - cite_ref-16

https://en.wikipedia.org/wiki/Root-knot_nematode - cite_ref-17

https://en.wikipedia.org/wiki/Root-knot_nematode - cite_ref-18

https://en.wikipedia.org/wiki/Root-knot_nematode - cite_ref-19

https://en.wikipedia.org/wiki/Root-knot_nematode - cite_ref-23

https://en.wikipedia.org/wiki/Root-knot_nematode - cite_ref-6Jatala, P. and bridge, J. 1990. Nematode parasites of root and tuber crops. In Plant parasitic nematodes in sub-tropical and tropical agriculture., pp 137-180. Luc, M., Sikora, R.A., Bridge, J., CABI Publishing, Wallingford, UK.

https://en.wikipedia.org/wiki/Root-knot_nematode - cite_ref-7Gapasin, R.M. (1980). Reaction of golden yellow cassava to Meloidogyne spp. Inoculation. *Annals of Tropical Research* 2:49-53.

https://en.wikipedia.org/wiki/Root-knot_nematode - cite_ref-9Theberge, R. L. (1985). Common African Pests and Diseases of cassava, Yam, Sweet Potato and Cocoyam. International Institute of Tropical Agriculture (IITA). *Ibadan, Nigeria* 107 p.

Hussey, R. S. and Grundler, F. M. W. (1998). Nematode parasitism of plants. In: The Physiology and Biochemistry of free-living and plant-parasitic nematodes. Perry, R. N. and Wright, D. J. (Eds), CABI Publishing, UK. pp 213 – 243.

Madulu, J. and Trudgill, D. L. (1994). Influence of temperature on *Meloidogyne javanica. Nematologica,* 40, 230 - 243.

Makumbi-kidza, N. N., Speijer and Sikora R. A. (2000). Effects of *Meloidogyne incognita* on Growth and Storage-Root Formation of Cassava (Manihot esculenta). *J Nematol.;* 32(4S): 475–477.

Moens, Maurice., Roland N Perry. and James L Starr (2009). "Meloidogyne Species: a Diverse Group of Novel and Important Plant Parasites." In Root-knot Nematodes, ed. Roland N Perry, Maurice Moens, and James L Starr, 1–17. Wallingford, UK: CABI Publishing.

Neal, J. C. (1889). The root-knot disease of the peach, orange and other plants in Florida due to the work of Anguillula. *Bull. I.S. Bur. Ent.*20.31pp.

Norton, D. C. and Niblack, T. L. (1991). Biology and ecology of nematodes. In: *Manual of Agricultural Nematology,* Nickle, W. R. (Ed), Marcel Dekker, New York. pp 47 – 68.

Pline and Dusenbery (1987). Responses of the plant-parasitic nematode *Meloidogyne incognita* to carbon dioxide determined by video camera-computer tracking. *J. Chem. Ecol.* 13: 873- 888.

Sijmons, P. C., Atkinson, H. J. and Wyss, U. (1994). Parasitic strategies of root nematodes and associated host cell responses. *Annual Review of Phytopathology*, 32, 235 - 259.

Stirling, G. R., Stanton, J. M. and Marshall, J. W. (1992). The importance of plant-parasitic nematodes to Australian and New Zealand agriculture. Australasian Plant Pathology, 21, 104 - 115.

Surdiman and Webster, J. M. (1995). Effect of ammonium ions on egg hatching and second-stage juveniles of Meloidogyne incognita in axenic tomato root culture. *Journal of Nematology*, 27, 346 - 352.

Trudgill, D. L. (1995). An assessment of the relevance of thermal time relationships to nematology. *Fundamental and Applied Nematology*, 18, 407 - 417.

Wood, W. B. (1988). Introduction to C.elegans. In::The Nematode Caenorhabditis elegans, W. B. Wood (Ed), Cold Spring Harbour Laboratory, New York. pp 1 – 16.

Cyst Forming Nematode

http://www.doacs.state.fl.us/pi/enpp/nema/nemacirc/nem158.pdf The cereal cyst nematode, *Heterodera avenae*]

Cetipedes

Barnes, Robert D. (1982). Invertebrate Zoology. Philadelphia, PA: Holt-Saunders International. pp. 810–816.ISBN 0-03-056747-5.

Bush, Sean P., King, Bradley O., Norris, Robert L. and Stockwell, Scott A. (2001). "Centipede envenomation".Wilderness and Environmental Medicine 12 (2): 93–99.doi:10.1580/1080- 6032(2001) 012[0093:CE] 2.0.CO; 2.PMID 11434497.

Edgecombe, G. D. and Giribet, G. (2002). "Myriapod phylogeny and the relationships of Chilopoda". In Bousquets, J. Llorente; Morrone, J. J. Biodiversidad, Taxonomía y Biogeografia de Artrópodos de México: Hacia una Síntesis de su Conocimiento, Volumen III. Universidad Nacional Autónoma de México. pp. 143–168.

Edgecombe, Gregory D. and Giribet, Gonzalo (2008). "A New Zealand species of the trans- Tasman centipede order Craterostigmomorpha (Arthropoda: Chilopoda) corroborated by molecular evidence". Invertebrate Systematics 22: 1–15. doi:10.1071/IS07036.

Fox, Richard (2006). "Invertebrate Anatomy OnLine: Scutigera coleoptrata, house centipede". Lander University.

Giribet, Gonzalo, and Edgecombe, Gregory D. (2006)."Conflict between datasets and phylogeny of centipedes: an analysis based on seven genes and morphology" (PDF).Proceedings of the Royal Society B 273 (1586): 531538.doi:10.1098/rspb.2005.3365. PMC 1560052.

https://en.wikipedia.org/wiki/Centipede - cite_ref-10Hölldobler, Bert *and* Wilson, Edward O. *(1990). "The specialized predators"*. The Ants. Harvard University Press. *pp. 557–572.* ISBN 978-0-674-04075-5.

https://en.wikipedia.org/wiki/Centipede - cite_ref-12*Einser, Thomas, Eisner, Maria and Siegler, Melody (2005).Secret weapons: defenses of insects, spiders, scorpions and other many-legged creatures. Cambridge, Mass.: The Belknap Press. pp. 33–36.* ISBN 978-0-674-01882-2.

https://en.wikipedia.org/wiki/Centipede - cite_ref-2*Arthur, W. (2002). "The interaction between developmental bias and natural selection from centipede segmentation to a general hypothesis". Heredity* ***89*** *(4): 239– 246.* doi:10.1038/sj.hdy.6800139. PMID 12242638.

https://en.wikipedia.org/wiki/Centipede - cite_ref-3*Arthur, Wallace and Chapman, Ariel D. (2005). "The centipede Strigamia maritima: what it can tell us about development and evolution of segmentation". BioEssays* **27***(6): 653– 660.* doi:10.1002/bies.20234.PMID 15892117.

https://en.wikipedia.org/wiki/Centipede - cite_ref-5*Adis, Joachim and Harvey, Mark S. (2000). "How many Arachnida and Myriapoda are there worldwide and in Amazonia?". Studies on Neotropical Fauna and Environment* **35** *(2): 139– 141.* doi:10.1076/0165-0521(200008)35:2;1-9;FT139.

https://en.wikipedia.org/wiki/Centipede - cite_ref-8*Molinari, Jesús, Gutiérrez, Eliécer E., de Ascenção, Antonio A., Nassar, Jafet M., Arends, Alexis and Márquez, Robert J. (2005).* "Predation by giant centipedes,Scolopendra gigantea, on three species of bats in a Venezuelan cave" *(PDF). Caribbean Journal of Science* **4***(2): 340– 346.*

Jacobs, Steven J. (2009). House Centipede (PDF). Pennsylvania State University. In 1902, C.L. Marlatt, an entomologist with the United States Department of Agriculture writes in Circular #48

Lewis, J. G. E. (2007). The Biology of Centipedes. Cambridge University Press. ISBN 978-0- 521-03411-1.

Lloyd, John (2006). The Book of General Ignorance. London: Bloomsbury House. p. 119. ISBN 0571273785. Retrieved June 10, 2014.

Minelli, Alessandro (2009). "Evolutionary developmental biology does not offer a significant challenge to the neo-Darwinian paradigm". In Ayala, Francisco J.; Arp, Robert.Contemporary Debates in Philosophy of Biology. John Wiley and Sons. pp. 213– 226. ISBN 978-1-4051-5999-9.

Zabs-Dkar Tshogs-Drug-Ran-Grol. (1994). The Life of Shabkar: *The Autobiography of a Tibetan Yogin.* Albany: SUNY Press,. p. 295.

Millipedes

Birkinshaw, Christopher R. (1999). "Use of Millipedes by Black Lemurs to Anoint Their Bodies". Folia Primatologica 70 (3): 170–171.doi:10.1159/000021691.

Blower, J. Gordon (1985).Millipedes: Keys and Notes for the Identification of the Species. London: Published for the Linnean Society of London and the Estuarine and Brackish- Water Sciences Association by E.J. Brill.ISBN 9004076980.

Brewer, Michael S., Sierwald, Petra, and Bond, Jason E. (2012)."Millipede Taxonomy after 250 Years: Classification and Taxonomic Practices in a Mega-Diverse yet Understudied Arthropod Group".PLoS ONE 7 (5): e37240.doi:10. 1371/journal. pone. 0037240 . PMC 3352885.PMID 22615951.

Capinera, John L. (2008). "Millipedes". Encyclopedia of Entomology. Springer. pp. 2935– 2397. ISBN 978-1-4020-6242-1.

F. (1998). "Colony composition and specialized predation on millipedes in the enigmatic ponerine ant genus Probolomyrmex (Hymenoptera, Formicidae)"*(PDF). Insectes Sociaux* ***45*** *(1): 79– 83.* doi:10.1007/s000400050070.

Golovatch, Sergei I., Kime. and R. Desmond (2009). "Millipede (Diplopoda) distributions: a review." (PDF). Soil Organisms **81**(3): 565–597.

Hoffman, Richard L. (2000)."Milliped or Millipede?" (PDF).Bulletin of the British Myriapod Group **16**: 36–37.

Hopkin, Stephen P. and Read, Helen J. (1992). The Biology of Millipedes. Oxford: Oxford University Press. ISBN 0198576994.

https://en.wikipedia.org/wiki/Millipede - cite_ref-11*Marek, Paul E. and Moore, Wendy (2015).* "Discovery of a glowing millipede in California and the gradual evolution of bioluminescence in Diplopoda".*Proceedings of the National Academy of Sciences* ***112*** *(20): 6419–6424.*doi:10.1073/pnas.1500014112.

https://en.wikipedia.org/wiki/Millipede - cite_ref-23*Drago, Leandro., Fusco, Giuseppe., Garollo, Elena and Minelli, Alessandro (2011). "Structural aspects of leg-to-gonopod metamorphosis in male helminthomorph millipedes (Diplopoda)". Frontiers in Zoology* ***8****(1): 19.* doi:10.1186/1742-9994-8-19.

https://en.wikipedia.org/wiki/Millipede - cite_ref-39*Forthman, M., and Weirauch, C. (2012).* "Toxic associations: A review of the predatory behaviors of millipede assassin bugs (Hemiptera: Reduviidae: Ectrichodiinae)"*(PDF). European Journal of Entomology* ***109*** *(2): 147–153.*doi:10.14411/eje.2012.019.

https://en.wikipedia.org/wiki/Millipede - cite_ref-41*Yasumasa Kuwahara; Hisashi Ômura and Tsutomu Tanabe (2002). "2-Nitroethenylbenzenes as natural products in millipede defense secretions".* Naturwissenschaften ***89****(7): 308– 310.*Bibcode: 2002 NW.89.308K.doi:10.1007/s00114-002-0328-9.PMID 12216861.

https://en.wikipedia.org/wiki/Millipede - cite_ref-46*Eisner, Thomas, Eisner, Maria and Deyrup, Mark (1996).* "Millipede defense: use of detachable bristles to entangle ants" (PDF*).Proceedings of the National Academy of Sciences* ***93*** *(20): 10848–10851.*Bibcode:1996PNAS.9310848E.doi: 10.1073/pnas. 93.20.10848. PMC 38244. PMID 8855269.

https://en.wikipedia.org/wiki/Millipede - cite_ref-55https://en.wikipedia.org/wiki/Millipede - cite_ref-56*Hudson, B. and Parsons, G. (1997). "Giant millipede 'burns' and the eye".Transactions of the Royal Society of Tropical Medicine and Hygiene* ***91****(2): 183–185.* doi:10.1016/S0035- 9203(97)90217-0.PMID 9196764.

https://en.wikipedia.org/wiki/Millipede - cite_ref-59*Ebregt, E., Struik, P. C., Odongo, B. and Abidin, P. E. (2005). "Pest damage in sweet potato, groundnut and maize in north-eastern Uganda with special reference to damage by millipedes (Diplopoda)". NJAS – Wageningen Journal of Life Sciences* **53** *(1): 49– 69.*doi:10.1016/S1573-5214(05)80010-7.

https://en.wikipedia.org/wiki/Millipede - cite_ref-60https://en.wikipedia.org/wiki/Millipede - cite_ref-Peckham2013_61-0*Peckham, Matt (2013).*"Millipedes – Yes, Millipedes – May Be Responsible for Australian Train Crash". *Time Newsfeed.* Time Magazine. Retrieved 31 October2013.

https://en.wikipedia.org/wiki/Millipede - cite_ref-63*Costa Neto, Eraldo M. (2007).* "The perception of Diplopoda (Arthropoda, Myriapoda) by the inhabitants of the county of Pedra Branca, Santa Teresinha, Bahia, Brazil". *Acta Biológica Colombiana* **12** *(2): 123–134.*

https://en.wikipedia.org/wiki/Millipede - cite_ref-66*Negi, C. S.; Palyal, V. S. (2007).*"Traditional Uses of Animal and Animal Products in Medicine and Rituals by the Shoka Tribes of District Pithoragarh, Uttaranchal, India" *(PDF). Studies on Ethno-Medicine* **1** *(1): 47–54.*

https://en.wikipedia.org/wiki/Millipede - cite_ref-71*Avirovik, Dragan., Butenhoff, Bryan and Priya, Shashank (2014). "Millipede-inspired locomotion through novel U-shaped piezoelectric motors". Smart Materials and Structures* **23** *(3): 037001.* doi:10.1088/0964-1726/23/3/037001.

https://en.wikipedia.org/wiki/Millipede - cite_ref-72 https://en.wikipedia.org/wiki/Millipede - cite_ref-73*Shelley, Rowland M. (1999).* "Centipedes and Millipedes with Emphasis on North American Fauna". *The Kansas School Naturalist* **45** *(3): 1–16.*

https://en.wikipedia.org/wiki/Millipede - cite_ref-7*Lockley, M. G. and Meyer, Christian (2013). "The tradition of tracking dinosaurs in Europe".* Dinosaur Tracks and Other Fossil Footprints of Europe. Columbia University Press. *pp. 25–52.* ISBN 9780231504607.

https://en.wikipedia.org/wiki/Millipede - cite_ref-Adis_1986_28-0https://en.wikipedia.org/wiki/Millipede - cite_ref-Burrows_et_al_1994_29-0*Burrows, F. J., Hales, D. F. and Beattie, A. J. (1994). "Aquatic millipedes in Australia: a biological enigma and a conservation saga".Australian Zoologist* **29** *(3–4): 213–216.* doi:10. 7882 /az.1994.007.

https://en.wikipedia.org/wiki/Millipede - cite_ref-Alagesan2005_57-0*Alagesan, P. and Muthukrishnan, J. (2005).* "Bioenergetics of the household pest, Xenobolus carnifex(Fabricius, 1775)" *(PDF).Peckiana* **4***: 3–14.*

https://en.wikipedia.org/wiki/Millipede - cite_ref-Baker2004_31-0*Barker, G. M. (2004). "Millipedes (Diplopoda) and Centipedes (Chilopoda)(Myriapoda) as predators* *of terrestrial gastropods". In Barker, G. M.* Natural enemies of terrestrial molluscs. *Wallingford, Oxfordshire, UK: CAB International. pp. 405–426.* ISBN 9780851990613.

https://en.wikipedia.org/wiki/Millipede - cite_ref-Barker_8-0*Barker, G.M. (2004).* Natural Enemies of Terrestrial Molluscs. *CABI. pp. 405–406.* ISBN 978-0- 85199-061-3.

https://en.wikipedia.org/wiki/Millipede - cite_ref-Biology_of_Centipedes_15-0*Lewis, J. G. E. (2008). The Biology of Centipedes (Digitally printed 1st paperback version. ed.). Cambridge:* Cambridge University Press. *pp. 110–111.*ISBN 9780521034111.

https://en.wikipedia.org/wiki/Millipede - cite_ref-Brewer.26Bond2013_85-0*Brewer, Michael S. and Bond, Jason E. (2013). "Ordinal-Level Phylogenomics of the Arthropod Class Diplopoda (Millipedes) Based on an Analysis of 221 Nuclear Protein-Coding Loci Generated Using Next-Generation Sequence Analyses". PLoS ONE **8** (11): e79935.* doi: 10.1371/journal.pone.0079935.

https://en.wikipedia.org/wiki/Millipede - cite_ref-Eisner_et_al_1998_34-0*Eisner, T., Eisner, M., Attygalle, A. B., Deyrup, M. and Meinwald, J. (1998).*"Rendering the inedible edible: circumvention of a millipede's chemical defense by a predaceous beetle larva. ". *Proceedings of the National Academy of Sciences of the United States of America* **95** *(3): 1108–13.*doi:10.1073/pnas.95.3.1108.PMID 9448293.

https://en.wikipedia.org/wiki/Millipede - cite_ref-Enghoff_cocoon_25-0*Enghoff, Henrik and Akkari, Nesrine (2011).* "A callipodidan cocoon (Diplopoda, Callipodida, Schizopetalidae)". *International Journal of Myriapodology* **5***: 49–53.*doi: 10.3897/ijm .5.1995.

https://en.wikipedia.org/wiki/Millipede - cite_ref-Enghoff2006invasion_58-0*Enghoff, Henrik and Kebapæi, Ümit (2008). "Calyptophyllum longiventre(Verhoeff, 1941) invading houses in Turkey, with the first description of the male (Diplopoda: Julida: Julidae)" . Journal of Natural History***42** *(31–32): 2143–2150.*doi: 10.1080/ 00222 930 8 0 2196055.

https://en.wikipedia.org/wiki/Millipede - cite_ref-Enghoffetal2014_68-0*Enghoff, Henrik, Manno, Nicola, Tchibozo, Sévérin, List, Manuela, Schwarzinger, Bettina, Schoefberger, Wolfgang, Schwarzinger, Clemens andPaoletti, Maurizio G. (2014).* "Millipedes as Food for Humans: Their Nutritional and Possible Antimalarial Value–A First Report". *Evidence-Based Complementary and Alternative Medicine* ***2014****: 1–9.*doi: 10.1155/2014/651768.

https://en.wikipedia.org/wiki/Millipede - cite_ref-Farfan.26Klompen2012_48-0*Farfan, Monica and Klompen, Hans (2012).* "Phoretic mite associates of millipedes (Diplopoda, Julidae) in the northern Atlantic region (North America, Europe)". *International Journal of Myriapodology* **7***: 69–91.*doi:10.3897/ijm.7.3064.

https://en.wikipedia.org/wiki/Millipede - cite_ref-Forgie_et_al_2002_37-0 *Forgie, Shaun A., Grebennikov, Vasily V. and Scholtz, Clarke H. (2002).*"Revision of Sceliages Westwood, a millipede-eating genus of southern African dung beetles

(Coleoptera: Scarabaeidae)" *(PDF).Invertebrate Systematics* **16** *(6): 931–955.* doi: 10.1071 /IT01025.

https://en.wikipedia.org/wiki/Millipede - cite_ref-Herbert2005_36-0*Herbert, D. G. (2000). "Dining on diplopods: remarkable feeding behaviour in chlamydephorid slugs (Mollusca: Gastropoda)". Journal of Zoology* **251** *(1): 1–5.*doi:10.1111/j.1469-7998. 2000.tb00586.x.

https://en.wikipedia.org/wiki/Millipede - cite_ref-Larsen_et_al_2009_38-0*Larsen, T. H; Lopera, A.; Forsyth, A.; Genier, F. (2009).* "From coprophagy to predation: a dung beetle that kills millipedes".*Biology Letters* **5** *(2): 152–155.*doi:10.1098/ rsbl.2008. 0654. PMC 2665820.PMID 19158030.

https://en.wikipedia.org/wiki/Millipede - cite_ref-Lawal.26Banjo2007_65-0*Lawal, O. A. and Banjo, A. D. (2007).*"Survey for the Usage of Arthropods in Traditional Medicine in Southwestern Nigeria" *(PDF).Journal of Entomology* **4** *(2): 104–112.* doi: 0.3923/je. 2007. 104.112.

https://en.wikipedia.org/wiki/Millipede - cite_ref-Linnaeus_76-0*Shelley, R. M. (2007).* "Taxonomy of extant Diplopoda (Millipeds) in the modern era: Perspectives for future advancements and observations on the global diplopod community (Arthropoda: Diplopoda)" *(PDF).Zootaxa* **1668***: 343–362.*

https://en.wikipedia.org/wiki/Millipede - cite_ref-Littoral_Myriapods_30-0*Barber, A. D. (2013).* "World Database of Littoral Myriapoda".World Register of Marine Species. Retrieved 25 October 2013.

https://en.wikipedia.org/wiki/Millipede - cite_ref-Marek_19-0*Marek, P., Shear, W. and Bond, J. (2012).* "A redescription of the leggiest animal, the millipedeIllacme plenipes, with notes on its natural history and biogeography (Diplopoda, Siphonophorida, Siphonorhinidae)". *ZooKeys* **241***(241): 77 112.*doi:10.3897/ zookeys.241. 3831. PMC 3559107. PMID 23372415.

https://en.wikipedia.org/wiki/Millipede - cite_ref-Roncadori_et_al_1985_45-0*Roncadori, R. W., Duffey, S. S. and Blum, M. S. (1985).* "Antifungal Activity of Defensive Secretions of Certain Millipedes". *Mycologia* **77***(2): 185–191.*doi:10.2307/3793067.

https://en.wikipedia.org/wiki/Millipede - cite_ref-Saporito_et_al_2003_33-0*Saporito, R. A., Donnelly, M. A., Hoffman, R. L., Garraffo, H. M. and Daly, J. W. (2003). "A Siphonotid Millipede (Rhinotus) as the Source of Spiropyrrolizidine Oximes of Dendrobatid Frogs". Journal of Chemical Ecology* **29** *(12): 2781–2786.*doi:10.1023/ B: JOEC. 0000 00 80 65.28364.a0.

https://en.wikipedia.org/wiki/Millipede - cite_ref-Shelley.26Golavatch Atlas2011_26-0 *Shelley, Rowland M.; Golavatch, and Sergei I. (2011). "Atlas of myriapod biogeography. I. Indigenous ordinal and supra-ordinal distributions in the Diplopoda: Perspectives on taxon origins and ages, and a hypothesis on the origin and early evolution of the class".Insecta Mundi* **158***: 1–134.*

https://en.wikipedia.org/wiki/Millipede - cite_ref-Stoev.26Bavel2005_47-0*Stoev, Pavel, and Lapeva-Gjonova, Albena (2005).* "Myriapods from ant nests in Bulgaria (Chilopoda, Diplopoda)" *(PDF). Peckiana* **4***: 131–142.*

https://en.wikipedia.org/wiki/Millipede - cite_ref-Stoev_et_al_2010_62-0*Stoev, Pavel; Zapparoli, Marzio; Golovatch, Sergei; Enghoff, Henrik; Akkari, Nesrine and Barber, Anthony (2010).* "Myriapods (Myriapoda). Chapter 7.2. In: Roques *et al.* (Eds). Alien terrestrial arthropods of Europe". *BIORISK – Biodiversity and Ecosystem Risk Assessment* **4**: *97–130.* doi:10.3897/biorisk.4.51.

https://en.wikipedia.org/wiki/Millipede - cite_ref-Swafford.26Bond2010_49-0https://en.wikipedia.org/wiki/Millipede - cite_ref-Mart.C3.ADnez-Torres_et_al_2011_50-0https://en.wikipedia.org/wiki/Millipede - cite_ref-MJA-Mason_52-0https://en.wikipedia.org/wiki/Millipede - cite_ref-53*Shpall, S. and Frieden, I. (1991). "Mahogany discoloration of the skin due to the defensive secretion of a millipede". Pediatric Dermatology 8(1): 25–27.* doi:10.1111/j.1525-1470.1991. tb00834.x. PMID 1862020.

https://en.wikipedia.org/wiki/Millipede - cite_ref-valederrama_43-0*Valderrama, Ximena; Robinson, John G., Attygalle, Athula B. and Eisner, Thomas (2000). "Seasonal anointment with millipedes in a wild primate: a chemical defense against insects". Journal of Chemical Ecology* **26** *(12): 2781–2790.*doi:10.1023/A:1026489826714.

https://en.wikipedia.org/wiki/Millipede - cite_ref-Wesener_et_al._2011_24-0*Wesener, Thomas; Köhler, Jörn; Fuchs, Stefan and van den Spiegel, Didier (2011). "How to uncoil your partner–"mating songs" in giant pill-millipedes (Diplopoda: Sphaerotheriida)". Naturwissenschaften* **98** *(11): 967–975.* doi:10.1007/s00114-011-0850-8.

Kraus, O. (2005). "On the structure and biology of Arthropleuraspecies (Atelocerata, Diplopoda; Upper Carboniferous/Lower Permian)". Verhandlungen des Naturwissenschaftlichen Vereins in Hamburg **41**: 5–23.

Martínez-Torres, Shirley Daniella; Daza, Álvaro Eduardo Flórez; Linares-Castillo and Edgar Leonardo (2011). "Meeting between kingdoms: discovery of a close association between Diplopoda and Bryophyta in a transitional Andean-Pacific forest in Colombia" . *International Journal of Myriapodology* **6**: *29–36.*doi:10.3897/ijm.6.2187.

Mason, G., Thompson, H., Fergin, P. and Anderson, R. (1994). "Spot diagnosis: the burning millipede".Medical Journal of Australia **160**(11): 718–726. PMID 8202008.

Minelli, Alessandro and Golovatch, Sergei I. (2001). "Myriapods". In Levin, Simon A. Encyclopedia of Biodiversity (PDF). pp. 291–303.ISBN 0122268652.

Russell Garwood and Gregory Edgecombe (2011). "Early terrestrial animals, evolution and uncertainty". Evolution, Education, and Outreach **4** (3): 489–501.doi:10.1007/s12052-011- 0357-y.

Shear, W. (2011). "Class Diplopoda de Blainville in Gervais, 1844. In: Zhang, Z.-Q. (Ed.) Animal biodiversity: An outline of higher-level classification and survey of taxonomic richness " *(PDF).Zootaxa* **3148**: *159–164.*

Shear, William A. and Edgecombe, Gregory D. (2010)."The geological record and phylogeny of the Myriapoda".*Arthropod Structure and Development* **39** *(2–3): 174–190.*doi:10. 1016/j. asd. 2009.11.002.PMID 19944188.

Sierwald, Petra and Bond, Jason E. (2007). *"Current Status of the Myriapod Class Diplopoda (Millipedes): Taxonomic Diversity and Phylogeny". Annual Review of Entomology* **52** *(1): 401–420.*doi:10.1146/annurev.ento.52.111805.090210. PMID 17163800.

Sues, Hans-Dieter (2011). "Largest Land-Dwelling "Bug" of All Time". National Geographic. Retrieved 25 February 2016.

Swafford, Lynn and Bond, Jason E. (2010). "Failure to cospeciate: an unsorted tale of millipedes and mites". Biological Journal of the Linnean Society **101** (2): 272–287.doi:10.1111/j.1095- 8312.2010.01499.x.

Weldon, Paul J., Aldich, Jeffrey R., Klun, Jerome A., Oliver, James E. and Debboun, Mustapha (2003). "Benzoquinones from millipedes deter mosquitoes and elicit self-anointing in capuchin monkeys (Cebus spp.)". Naturwissenschaften **90** (7): 301–305.Bibcode:2003 NW.90.301W. doi:10. 1007/s00114-003-0427-2.PMID 12883771.

Wilson, Heather M. and Anderson, Lyall I. (2004). *"Morphology and taxonomy of Paleozoic millipedes (Diplopoda: Chilognatha: Archipolypoda) from Scotland".Journal of Paleontology* **78** *(1): 169–184.* doi:10.1666/0022-3360(2004)078<0169: MATOPM> 2.0.CO;2.

Snails

De Vries, Ad (1976). Dictionary of Symbols and Imagery. Amsterdam: North-Holland Publishing Company. p. 430. ISBN 0-7204-8021-3.

Fredericks, Anthony D. (2010). How Long Things Live. *Stackpole Books. p. 73.* ISBN 9780811736220. Retrieved19 June 2012.

https://en.wikipedia.org/wiki/Snail - cite_ref-2https://en.wikipedia.org/wiki/Snail - cite_ref-3*Brieva, A., Philips, N., Tejedor, R., Guerrero, A., Pivel, J. P., Alonso-Lebrero, J. L. and Gonzalez, S. (2008).* "Molecular basis for the regenerative properties of a secretion of the mollusk Cryptomphalus aspersa". *Skin Pharmacology and Physiology***21** *(1): 15–21.* doi: 10. 1159/000109084. ISSN 1660-5527.

https://en.wikipedia.org/wiki/Snail - cite_ref-5https://en.wikipedia.org/wiki/Snail - cite_ref-Cooper92_6-0*Cooper, J.C. (1992). Symbolic and Mythological Animals. London: Aquarian Press. p. 213.* ISBN 1-85538-118-4.

Rodents

Rattus rattus

Amori, G., Hutterer, R., Kryštufek, B., Yigit, N., Mitsain, G. and Palomo, L.J. (2008). *"Rattus rattus"*. IUCN Red List of Threatened Species. *Version 2011.1.* International Union for Conservation of Nature. Retrieved 9 August 2011.

Bennet, Stuart M.(2011). "The Black Rat (*Rattus Rattus*)".*The Pied Piper*. Retrieved 22 April 2011.

https://en.wikipedia.org/wiki/Black_rat - cite_ref-14https://en.wikipedia.org/wiki/Black_rat - cite_ref-15*Dowding, J.E. and Murphy, EC (1994).* "Ecology of

Ship Rats (*Rattus rattus*) in a Kauri (Agathis australis) Forest in Northland, New Zealand" *(PDF). New Zealand Journal of Ecology* ***18*** *(1): 19–28.* ISSN 0110-6465.

https://en.wikipedia.org/wiki/Black_rat - cite_ref-4https://en.wikipedia.org/wiki/Black_rat - cite_ref-5Alderton, David (1996). Rodents of the World, p. 29.ISBN 0-8160-3229-7

https://en.wikipedia.org/wiki/Black_rat - cite_ref-Cox_17-0*Cox, MPG; Dickman, C.R. and Cox, WG (2000). "Use of habitat by the black rat (Rattus rattus) at North Head, New South Wales: an observational and experimental study".Austral Ecology* ***25*** *(4): 375–85.* doi:10.1046/j.1442-9993.2000.01050.x.

https://en.wikipedia.org/wiki/Black_rat - cite_ref-Rackham_6-0 "Rattus rattus – Roof rat". *Wildlife Information Network*. Retrieved 22 April 2011.

Marsh, Rex E. (1994). "Roof Rats". *Internet Center for Wildlife Damage Management. Prevention and Control of Wildlife Damage*. Retrieved 22 April 2011.

McCormick, M. (2003). "Rats, Communications, and Plague: Toward an Ecological History" *(PDF).Journal of Interdisciplinary History* ***34*** *(1): 1–25.*doi: 10.1162/0022 1950 3322645439. ISSN 0022-1953.

Schwartz, Charles Walsh and Schwartz, Elizabeth Reeder (2001). The Wild Mammals of Missouri, University of Missouri Press, ISBN 978-0-8262-1359-4, p. 250.

Rattus norvegicus

https://en.wikipedia.org/wiki/Brown_rat - cite_ref-10*Clark, B. R. and Price, E. O. (1981). "Sexual maturation and fecundity of wild and domestic Norway rats (Rattus norvegicus)". Journal of Reproduction and Fertility* ***63*** *(1): 215– 220.* doi: 10.1530/jrf.0.0630215.

https://en.wikipedia.org/wiki/Brown_rat - cite_ref-12*Hanson, Anne (2007).* "What Do Rats See?".*Rat Behavior and Biology. ratbehavior.org*. Retrieved1 December 2019.

https://en.wikipedia.org/wiki/Brown_rat - cite_ref-13https://en.wikipedia.org/wiki/Brown_rat - cite_ref-14 *Smith, J. David; Beran, M. J., Couchman, J. J. and Coutinho, M. V. C. (2008).* "The Comparative Study of Metacognition: Sharper Paradigms, Safer Inferences" *(PDF).Psychonomic Bulletin and Review* ***15*** *(4): 679–691.*doi:10.3758/pbr.15.4.679.

https://en.wikipedia.org/wiki/Brown_rat - cite_ref-15*Brunelli, Susan A., Shair, Harry N. and Hofer, Myron A. (1994). "Hypothermic vocalizations of rat pups (Rattus norvegicus) elicit and direct maternal search behavior".Journal of Comparative Psychology* ***108*** *(3): 298–303.*doi:10.1037/0735-7036.108.3.298.

https://en.wikipedia.org/wiki/Brown_rat - cite_ref-3 Donaldson, Henry Herbert. (1915) The Rat. pp. 13.

https://en.wikipedia.org/wiki/Brown_rat - cite_ref-4https://en.wikipedia.org/wiki/Brown_rat - cite_ref-5https://en.wikipedia.org/wiki/Brown_rat - cite_ref-6https://en.wikipedia.org/wiki/Brown_rat - cite_ref-7

https://en.wikipedia.org/wiki/Brown_rat - cite_ref-8https://en.wikipedia.org/wiki/Brown_rat - cite_ref-9"Pest info- Rodents". *BASF- The Chemical Company*. Retrieved 14 January 2014.

Ruedas, L. (2008). *"Rattus norvegicus"*. IUCN Red List of Threatened Species. *Version 2011.1.* International Union for Conservation of Nature. Retrieved 9 August 2011.

Mus musculus

Berry, R.J. (1970). "The natural history of the house mouse" *(PDF). Field Studies (Field Studies Council)* **3***: 219–62.* Retrieved 18 December 2013.

Dobson, F Stephen and Baudoin, Claude (2002). "Experimental tests of spatial association and kinship in monogamous mice (Mus spicilegus) and polygynous mice (Mus musculus domesticus)". Canadian Journal of Zoology,**80** (6): 980–6. doi:10.1139/z02-055.

Frynta, Daniel; Slábová, Markéta; Váchová, Hana; Volfová, Radka and Munclinger, Pavel (2005). "Aggression and commensalism in house mouse: A comparative study across Europe and the near east".Aggressive Behavior, **31** (3): 283–93.doi:10.1002/ab.15555.

Gray, Samantha J and Hurst, Jane L. (1997). "Behavioural mechanisms underlying the spatial dispersion of commensal Mus domesticusand grassland Mus spretus".Animal Behaviour, **53** (3): 511–24.doi:10.1006/anbe.1996.0301.

https://en.wikipedia.org/wiki/House_mouse - cite_ref-11*Greene, Eunice Chace (1935). Anatomy of the Rat. Transactions of the American Philosophical Society* **27**.JSTOR 1005513. OCLC 685221899.

https://en.wikipedia.org/wiki/House_mouse - cite_ref-13*Buck, C. W., Tolman, N. and Tolman, W. (November 1925). "The Tail as a Balancing Organ in Mice". Journal of Mammalogy* **6** *(4): 267–71.* doi:10.2307/1373415.JSTOR 1373415.

https://en.wikipedia.org/wiki/House_mouse - cite_ref-15 *Drickamer, Lee C. (2005).* "Use of the tail for communication in house mice". *In Sánchez-Cordero, Víctor; Medellín, Rodrigo A. Contribuciones mastozoológicas en homenaje a Bernardo Villa [Mammal Collection in Honor of Bernardo Villa] (in Spanish). UNAM. pp. 157– 62.* ISBN 978-970-32-2603-0.

https://en.wikipedia.org/wiki/House_mouse - cite_ref-17*Musser, Guy G. and Carleton, Michael D. (2005).*"Superfamily Muroidea". *In Wilson, Don E.; Reeder, DeeAnn M.* Mammal Species of the World: A Taxonomic and Geographic Reference *(3rd ed.). Baltimore: Johns Hopkins University Press. pp. 894–1531.* ISBN 978- 0-8018-8221-0.

https://en.wikipedia.org/wiki/House_mouse - cite_ref-19

https://en.wikipedia.org/wiki/House_mouse - cite_ref-20*Hilscher-Conklin, Caryl (1998). "Rattus Biologicus*: Coprophagy: Healthy Behavior For Your Rats". *Rat and Mouse Gazette.*

https://en.wikipedia.org/wiki/House_mouse - cite_ref-21*Horn, Charles C.; Kimball, Bruce A.; Wang, Hong; Kaus, James; Dienel, Samuel; Nagy, Allysa; Gathright, Gordon R.; Yates, Bill J. and Andrews, Paul L. R. (2013).* "Why Can't Rodents Vomit? A Comparative Behavioral, Anatomical, and Physiological Study". *PLoS ONE 8 (4):e60537.*Bibcode:2013PLoSO.860537H.doi:10.1371/journal.pone.0060537. PM C 3622671.PMID 23593236. Lay summary – *Live Science (April 16, 2013).*

https://en.wikipedia.org/wiki/House_mouse - cite_ref-22*Tattersall, Françoise H. and Smith, R. H.; Nowell, F. (1997).*"Experimental colonisation of contrasting habitats by house mice". *Zeitschrift für Säugetierkunde,**62** (6): 350–8.*

https://en.wikipedia.org/wiki/House_mouse - cite_ref-23*Moro, Dorian and Morris, Keith (2000). "Movements and refugia of Lakeland Downs short-tailed mice, Leggadina lakedownensis, and house mice, Mus domesticus, on Thevenard Island, Western Australia". Wildlife Research,* **27***(1): 11–20.* doi:10.1071/WR99016.

https://en.wikipedia.org/wiki/House_mouse - cite_ref-26 *Wolff, Robert J. (2009). "Mating behaviour and female choice: Their relation to social structure in wild caught House mice (Mus musculus) housed in a semi-natural environment". Journal of Zoology,* **207***: 43–51.*doi:10.1111/j.1469-7998.1985.tb04914.x.

https://en.wikipedia.org/wiki/House_mouse - cite_ref-29*Gerlach, Gabriele (1996). "Emigration mechanisms in feral house mice - a laboratory investigation of the influence of social structure, population density, and aggression".Behavioral Ecology and Sociobiology,* **39** *(3): 159–70.*doi:10.1007/s002650050277. JSTOR 4601248.

https://en.wikipedia.org/wiki/House_mouse - cite_ref-6*Baker R.O., Bodman G.R. and Timm RM (1994).* "Rodent-Proof Construction and Exclusion Methods". *In Hygnstrom SE, Timm RM, Larson GE. Prevention and Control of Wildlife Damage. University of Nebraska-Lincoln.*

https://en.wikipedia.org/wiki/House_mouse - cite_ref-9*Hotchkiss, A. K. and Vandenbergh, J. G. (2005). "The anogenital distance index of mice (Mus musculus domesticus): An analysis". Contemporary Topics in Laboratory Animal Science,* **44** *(4): 46–8.* PMID 16050669.

https://en.wikipedia.org/wiki/House_mouse - cite_ref-Terszowski_16-0*Terszowski, G., Müller, Susanna M., Bleul, Conrad C., Blum, Carmen; Schirmbeck, Reinhold; Reimann, Jörg; Du Pasquier, Louis; Amagai, Takashi; Boehm, Thomas; Rodewald and Hans- Reimer (2006). "Evidence for a Functional Second Thymus in Mice" . Science* **312** *(5771): 284–7.*Bibcode: 2006Sci. 312.284T. doi:10. 1126/science. 1123497. PMID 16513945.

Musser G., Amori G., Hutterer R., Kryštufek B., Yigit N. and Mitsain G (2008). Mus musculus. In: IUCN 2008. IUCN Red List of Threatened Species. Retrieved 10 October 2008.

Odling Smee, Lucy (2007). *"Mice made to see a rainbow of colours". News @nature.* doi: 10. 1038/news070319-12.

Prager, E. M., Orrego, C. and Sage, R. D. (1998). "Genetic variation and phylogeography of central Asian and other house mice, including a major new mitochondrial lineage in Yemen" . *Genetics* **150** *(2): 835–61.* PMC 1460354.PMID 9755213.

Siegel, Michael I. (1970). *"The tail, locomotion and balance in mice". American Journal of Physical Anthropology,* **33***: 101–2.* doi:10.1002/ajpa.1330330113.

Szenczi, Peter; Bánszegi, Oxana; Groó, Zita and Altbäcker, Vilmos (2012). "Development of the Social Behavior of Two Mice Species with Contrasting Social Systems". Aggressive Behavior, **38** (4): 288–97.doi:10.1002/ab.21431. PMID 25363698.

Bandicota bengalensis

https://en.wikipedia.org/wiki/Lesser_bandicoot_rat - cite_ref-2 "Bandicoot (Rat)". *Britannica encyclopedia*. Retrieved 17 May 2012.

https://en.wikipedia.org/wiki/Lesser_bandicoot_rat - cite_ref-5 *Dhar, P. and Singla, N. (2014). "Effect of triptolide on reproduction of female lesser bandicoot rat, Bandicota bengalensis.".Drug Chem Toxicol* **37** *(4): 448–58.* doi: 10.3109/014805 45.2014. 884111. PMID 24490988.

https://en.wikipedia.org/wiki/Lesser_bandicoot_rat - cite_ref-6*Brooks, J.E. and Htun, P.T. (1980).* "Laboratory evaluation of scilliroside used as a rodenticide against the lesser bandicoot rat, B*andicota bengalensis". J Hyg (Lond)* **85***: 227–34.* doi: 10.1017/s 00 22172400063269. PMC 2133936. PMID 7451959

Bandicota indica

Baillie J. (1996). *"Bandicota indica"*. 2006 IUCN Red List of Threatened Species. Downloaded on 19 July 2007.

https://en.wikipedia.org/wiki/Greater_bandicoot_rat - cite_ref-5*Singh, K. I., Krishnasamy, M. and Ambu, S. (1992). "The large bandicoot rat, Bandicota indica, a new host for Schistosoma spindale, Montgomery, 1906, in Peninsular Malaysia". The Southeast Asian journal of tropical medicine and public health* **23** *(3): 537– 538.* PMID 1488714

https://en.wikipedia.org/wiki/Greater_bandicoot_rat - cite_ref-Singh_1997_4-0 *Inder Singh, K., Krishnasamy, M., Ambu, S., Rasul, R. and Chong, N. L. (1997). "Studies on animal schistosomes in Peninsular Malaysia: Record of naturally infected animals and additional hosts of Schistosoma spindale". The Southeast Asian journal of tropical medicine and public health,* **28** *(2): 303–307.* PMID 9444010.

Yapa, A. and Ratnavira, G. (2013). Mammals of Sri Lanka. Colombo: Field Ornithology Group of Sri Lanka. p. 1012. ISBN 978-955-8576-32-8.

Tatera indica

B. Kryštufek, G. Shenbrot, M. Sozen and S. Molur (2008). *"Tatera indica"*. IUCN Red List of Threatened Species.*Version 2011.2.* International Union for Conservation of Nature. Retrieved June 10, 2012.

Musser, G. G. and M. D. Carleton. (2005). Superfamily Muroidea. pp. 894–1531 in Mammal Species of the World: a Taxonomic and Geographic Reference. D. E. Wilson and D. M. Reeder eds. Johns Hopkins University Press, Baltimore.

Meriones hurrianae

Corbet, Gordon Barclay and John Edwards Hill (1992). The mammals of the Indomalayan region: a systematic review. New York: Oxford University Press. ISBN 0-19-854693-9. *McKenna, Malcolm C.; Susan K.Bell (1997). Classification of Mammals above the Species Level. New York: Columbia University Press.* ISBN 0-231-11012-X.

https://en.wikipedia.org/wiki/Meriones_(rodent) - cite_ref-3 Pavlinov, I. Ya., Yu. A. Dubrovskiy., O. L. Rossolimo and E.G. Potapova. 1990. Gerbils of the world. Moscow: Nauka. *Tong, Haiyan (1989). "Origine et évolution des Gerbillidae*

(Mammalia, Rodentia) en Afrique du Nord". Mémoires de la Société Géologique de France. Nouvelle Série **155**: *1–120.* ISBN 2-85363-050-1.

https://en.wikipedia.org/wiki/Meriones_(rodent) - cite_ref-4*Musser, Guy G. and Michael D. Carleton (2005). "Superfamily Muroidea". In Don E. Wilson and DeeAnn M. Reeder (eds.). Mammal species of the world: a taxonomic and geographic reference (3rd ed.). Baltimore: Johns Hopkins University Press. pp. 894–1531.*ISBN 0- 8018-8221-4.

Nowak, Ronald M. (1999). Walker's Mammals of the World, volume 2. London: Johns Hopkins University Press. ISBN 0-8018-5789-9.

Mus budoga

Baillie, J. (1996). *Mus booduga.* 2006 IUCN Red List of Threatened Species.Downloaded on 19 July 2007.

Musser, G.G. *and Carleton, M.D. (2005).* "Superfamily Muroidea". *In Wilson, D.E.; Reeder, D.M.* Mammal Species of the World: A Taxonomic and Geographic Reference *(3rd ed.). Johns Hopkins University Press. p. 1390.*ISBN 978-0-8018-8221-0. OCLC 62265494.

Indian crested porcupine/Indian porcupine *Hystrix indica*

Alkon, P. U. and Saltz, D. (1985). "Potatoes and the Nutritional Ecology of Crested Porcupines in a Desert Biome".*Journal of Applied Ecology,* **22** *(3): 727–737.* doi:10.2307/2403225.https://en.wikipedia.org/wiki/Indian_crested_porcupine - cite_ref-10

Alkon, Philip U. and Mitrani, David Saltz (1988). "Influence of Season and Moonlight on Temporal-Activity Patterns of Indian Crested Porcupines (Hystrix indica)". *Journal of Mammalogy,* **69** *(1): 71–80.* doi:10.2307/1381749. ISSN 0022-2372.

Alkon, Philip U. and Saltz, David (1988). "Foraging Time and the Northern Range Limits of Indian Crested Porcupines (Hystrix indica Kerr)". *Journal of Biogeography,* **15** *(3): 403– 408.* doi:10.2307/2845271.

Amori, G., Hutterer, R., Kryštufek, B., Yigit, N., Mitsain, G. and Muñoz, L. J. P. (2008). *Hystrix indica.* In: IUCN 2008. IUCN Red List of Threatened Species. Retrieved 19 November 2015.

Arslan, Atilla (2008). "On the Indian crested porcupine, Hystrix indica (Kerr, 1792) in Turkey". Pakistan Journal of Biological Sciences, **11**: *315–317.*

Gurung, K.K.; Singh, R. (1996). Field Guide to the Mammals of the Indian Subcontinent. San Diego: Academic Press.

https://en.wikipedia.org/wiki/Indian_crested_porcupine - cite_ref-:4_7-2 *Alkon, Philip U., Degen, A. Allan; Cohen, Anat and Pollak, Haya (1986).* "Seasonal Energy Requirements and Water Intakes of Indian Crested Porcupines (Hystrix indica) in Captivity" . *Journal of Mammalogy,* **67** *(2): 333–342.* doi:10.2307/1380887.

https://en.wikipedia.org/wiki/Indian_crested_porcupine - cite_ref-8*Gorbunov, A.V. (1985). "Features of the ecology of porcupines in the deserts of eastern Prikaspia". Soviet Journal of Ecology,***16**: *248–253.*

https://en.wikipedia.org/wiki/Indian_crested_porcupine - cite_ref-11*Gutterman, Yitzchak and Herr, Nir (1981).* "Influences of porcupine (Hystrix indica) activity on the slopes on the northern Negev mountains – Germination and vegetation renewal in different geomorphological types and slope directions". *Oecologia,***51** *(3): 332 334.* doi: 10. 1007/BF 00 540902. ISSN 0029-8549.

https://en.wikipedia.org/wiki/Indian_crested_porcupine - cite_ref-16*Hafeez, Shahid; S., Khan, G., Ashfaq, Muhammad; H. and Khan, Z (2015).* www.cabdirect.org. Retrieved 2015-11-20.

https://en.wikipedia.org/wiki/Indian_crested_porcupine - cite_ref-13*Hanley, Thomas A. (1982).* "The Nutritional Basis for Food Selection by Ungulates". *Journal of Range Management,* **35***(2): 146–151.* doi:10.2307/3898379.

https://en.wikipedia.org/wiki/Indian_crested_porcupine - cite_ref-12*Kadhim, Abdul-Hassain H. (1997).* "Distribution and reproduction of the Indian Crested Porcupine *Hystrix indica* (Hystricidae: Rodentia) in Iraq". *Zoology in the Middle East,* **15** *(1): 9– 12.* doi:10.1080/09397140.1997.10637731. ISSN 0939-7140.

https://en.wikipedia.org/wiki/Indian_crested_porcupine - cite_ref-21*Owens, Mark and Owens, Delia (1984).* Cry of the Kalahari. *Houghton Mifflin Harcourt* . ISBN 0395647800.

https://en.wikipedia.org/wiki/Indian_crested_porcupine - cite_ref-14 https://en.wikipedia.org/wiki/Indian_crested_porcupine - cite_ref-15*Yair, A. and Shachak, M. (1982).* "A case study of energy, water and soil flow chains in an arid ecosystem". *Oecologia,* **54***(3): 389–397.* doi:10.1007/BF00380008. ISSN 0029-8549.

Olsvig-Whittaker, L., Shachak, M. and Yair, A. (1983). "Vegetation patterns related to environmental factors in a Negev Desert watershed". *Vegetatio,* **54** *(3): 153–165.* doi: 10.1007/BF 000 47104. ISSN 0042-3106.

Prater, Stanley Henry (1965). The Book of Indian Animals. Bombay: Diocesan Press.

van Aarde, R. J. (1985). "Reproduction in captive female Cape porcupines (Hystrix africaeaustralis)". Journal of Reproduction and Fertility, **75***: 577–582.*

Rodent Management

https://en.wikipedia.org/wiki/Pest_control - cite_ref-sterile_9-0 *Alan M. Bowerman and Joe E. Brooks (1971). "Evaluation of U-5897 as a male chemosterilant for rat control". Journal of Wildlife Management,* **35** *(4): 618–624.* doi: 10.2307/3799765 . JSTOR3799765.

https://en.wikipedia.org/wiki/Rodenticide - cite_ref-3 *Meerburg B.G., Brom F.W. and Kijlstra A. (2008).* "The ethics of rodent control". *Pest Manag Sci,* **64** *(12): 1205–11.*doi:10.1002/ps.1623. PMID 18642329.

https://en.wikipedia.org/wiki/Rodenticide - cite_ref-pmid17655702_5-0*Kotsaftis P., Girtovitis F., Boutou A., Ntaios G. and Makris P.E. (2007).* "Haemarthrosis after superwarfarin poisoning". *Eur. J. Haematol.* **79** *(3): 255–7.*doi:10.1111/j.1600-0609.2007. 00904.x. PMID 17655702.

https://en.wikipedia.org/wiki/Rodenticide - cite_ref-pmid19000263_4-0*Vandenbroucke V., Bousquet-Melou A., De Backer P. and Croubels S (2008).* "Pharmacokinetics of eight anticoagulant rodenticides in mice after single oral administration". *J. Vet. Pharmacol. Ther,* **31** *(5): 437–45.*doi:10.1111/j.1365-2885.2008.00979.x.PMID 19000263.

Tracy Swartz (2014). "CTA to put rats on birth control". *Chicago Tribune.* Retrieved 10 January 2015.

Birds

Myna

*Ali, Salim and Ripley, S. Dillon (2001).Handbook of the Birds of India and Pakistan, Volume 5 (2 (paperback) ed.). India: Oxford University Press. p. 278.*ISBN 0-19-565938-4.

BirdLife International *(2012). "Acridotheres tristis".*IUCN Red List of Threatened Species. *Version 2013.2.*International Union for Conservation of Nature. Retrieved26 November 2013.

Choudhury A. (1998). "Common Myna feeding a fledgling koel". Journal of the Bombay Natural History Society, **95** *(1): 115.*

https://en.wikipedia.org/wiki/Common_myna - cite_ref-11*Griffin, Andrea S. (2008). "Social learning in Indian mynahs, Acridotheres tristis: the role of distress calls.".Animal Behaviour,* **75** *(1): 79–89.*doi:10.1016/j.anbehav.2007.04.008.

https://en.wikipedia.org/wiki/Common_myna - cite_ref-13 https://en.wikipedia.org/wiki/Common_myna - cite_ref-14*Mahabal, Anil (1991).* "Activity-time budget of Indian MynaAcridotheres tristis (Linnaeus) during the breeding season". *Journal of the Bombay Natural History Society(Bombay Natural History Society),* **90** *(1): 96–97.* Retrieved22 January 2011.

https://en.wikipedia.org/wiki/Common_myna - cite_ref-15*Pande, Satish; Tambe, Saleel; Clement, Francis M. and Sant, Niranjan (2003). Birds of Western Ghats, Kokan and Malabar (including birds of Goa). Mumbai: Bombay Natural History Society and Oxford University Press. pp. 312, 377.*ISBN 0-19-566878-2.

https://en.wikipedia.org/wiki/Common_myna - cite_ref-rasmussen_5-0*Rasmussen, P.C. and J.C. Anderton (2005). Birds of South Asia: The Ripley Guide. Vol 2. Smithsonian Institution and Lynx Edicions. p. 584.*

Mahabal, Anil and Vaidya, V.G. (1989). "Diurnal rhythms and seasonal changes in the roosting behaviour of Indian Myna Acridotheres tristis (Linnaeus)". *Proceedings of Indian Academy of Sciences (Animal Science) (Indian Academy of Sciences, Bangaluru),* **98** *(3): 199– 209.*doi:10.1007/BF03179646. Retrieved 22 January 2011.

Pande, Satish (2009). Latin names of Indian birds explained. Mumbai: Bombay Natural History Society and Oxford University Press. pp. 417, 506. ISBN 978-0-19-806625-5.

Rasmussen, Pamela C. and Anderton, John C. (2005).Birds of South Asia - The Ripley Guide (volume 2). Smithsonian Institution, Washington and Lynx edicions, Barcelona. pp. 584, 683. ISBN 84-87334-66-0.

Sparrow

BirdLife International *(2013). "Passer domesticus"*. IUCN Red List of Threatened Species. *Version 2013.2.* International Union for Conservation of Nature. Retrieved 12 March 2014.

*Groschupf, Kathleen (2001). "Old World Sparrows". In Elphick, Chris; Dunning, John B., Jr.; Sibley, David. The Sibley Guide to Bird Life and Behaviour. London: Christopher Helm. pp. 562–564.*ISBN 0-7136-6250-6.

https://en.wikipedia.org/wiki/House_sparrow - cite_ref-9 *Felemban, Hassan M. (1997).* "Morphological differences among populations of house sparrows from different altitudes in Saudi Arabia" *(PDF). The Wilson Bulletin,* ***109*** *(3): 539–544.*

Johnston, Richard F.; Selander, Robert K (1973). "Evolution in the House Sparrow. III. Variation in Size and Sexual Dimorphism in Europe and North and South America". The American Naturalist, ***107*** *(955): 373–390.* doi:10.1086/282841. JSTOR 2459538.

Peacock

Ali, S. and Ripley, S D (1980). Handbook of the birds of India and Pakistan ***2*** *(2 ed.). Oxford University Press. pp. 123–126.* ISBN 0-19-562063-1.

Baker, E.C.S. (1928). The Fauna of British India, Including Ceylon and Burma. *Birds.* Volume 5 *(2 ed.). Taylor and Francis, London. pp. 282–284.*

Bird Life International *(2012). "Pavo cristatus"*. IUCN Red List of Threatened Species. *Version 2013.2.* International Union for Conservation of Nature. Retrieved 26 November 2013.

https://en.wikipedia.org/wiki/Indian_peafowl - cite_ref-11*Sclater PL (1860).* "On the black-shouldered peafowl of Latham (*Pavo nigripennis*)". *Proc. Zool. Soc. London: 221–222.*

https://en.wikipedia.org/wiki/Indian_peafowl - cite_ref-12*Seth-Smith, D (1940). "Peafowl". Avicultural Magazine* ***5****: 205–206.*

https://en.wikipedia.org/wiki/Indian_peafowl - cite_ref-13*Somes, R.G. Jr. and R. E. Burger (1991). "Plumage Color Inheritance of the Indian Blue Peafowl (Pavo cristatus): Blue, Black-Shouldered, Cameo, and Oaten". Journal of Heredity,* ***82****: 64– 68.* doi:10.1093/jhered/82.1.64.

https://en.wikipedia.org/wiki/Indian_peafowl - cite_ref-2 *Weekley, E (1921).* An etymological dictionary of modern English. *John Murray, London* . ISBN 1-176-40695-7.

https://en.wikipedia.org/wiki/Indian_peafowl - cite_ref-4*Burton, R F (1884).* The book of the sword. *Chatto and Windus, London. p. 155.* ISBN 0-486- 25434-8.

https://en.wikipedia.org/wiki/Indian_peafowl - cite_ref-5*Whistler, Hugh (1949).* Popular handbook of Indian birds *(4 ed.). Gurney and Jackson, London. pp. 401–410.* ISBN 1-4067-4576-6.

https://en.wikipedia.org/wiki/Indian_peafowl - cite_ref-fbi_7-0*Blanford, W.T. (1898).* The Fauna of British India, Including Ceylon and Burma. Birds **4**. *Taylor and Francis, London. pp. 681–70.*

Johnsingh, A.J.T. and Murali, S. (1978). "The ecology and behaviour of the Indian Peafowl (Pavo cristatus) Linn. of Injar". J. Bombay Nat. Hist. Soc., ***75*** *(4): 1069–1079.*

Lal, Krishna (2007). Peacock in Indian art, thought and literature. *Abhinav Publications. pp. 11, 26, 139.* ISBN 81-7017-429-5.

Parrot

BirdLife International *(2012). "Psittacula krameri"*.IUCN Red List of Threatened Species. *Version 2013.2.*International Union for Conservation of Nature. Retrieved26 November 2013.

https://en.wikipedia.org/wiki/Rose-ringed_parakeet - cite_ref-2*Jobling, J. A. (1991). A Dictionary of Scientific Bird Names.* OUP. ISBN 0-19-854634-3.

https://en.wikipedia.org/wiki/Rose-ringed_parakeet - cite_ref-3Sailaja, R., Kotak, V. C., Sharp, P. J., Schmedemann, R. and Haase, E. (1988). Environmental, dietary, and hormonal factors in the regulation of seasonal breeding in free-living female Indian rose-ringed parakeets (*Psittacula krameri*). *Hormones and Behavior*, 22 (4): 518– 527.

https://en.wikipedia.org/wiki/Rose-ringed_parakeet - cite_ref-4Krishnaprasadan, T. N., Kotak, V. C., Sharp, P. J., Schmedemann, R. and Haase, E. (1988) . Environmental and hormonal factors in seasonal breeding in free-living male Indian rose- ringed parakeets (Psittacula krameri). *Hormones and Behavior*, 22 (4): 488–496.

https://en.wikipedia.org/wiki/Rose-ringed_parakeet - cite_ref-8 *London Bird Report (2006). London Natural History Society. 2006. p. 93.* ISBN 0-901009-22-9.

https://en.wikipedia.org/wiki/Rose-ringed_parakeet - cite_ref-DA_2003_5-0Alderton, David *(2003). The Ultimate Encyclopedia of Caged and Aviary Birds. London, England: Hermes House. pp. 189–190.* ISBN 1-84309-164-X.

Pigeon

Blechman, Andrew (2007). Pigeons-The fascinating saga of the world's most revered and reviled bird. *St Lucia, Queensland: University of Queensland Press.*ISBN 978-0-7022-3641-9.

https://en.wikipedia.org/wiki/Domestic_pigeon - cite_ref-6*Viguier, C. (1882). "Le sens de l'orientation et ses organes chez les animaux et chez l'homme". Rev. phil. France Etranger* ***14****: 1–36.*

https://en.wikipedia.org/wiki/Domestic_pigeon - cite_ref-5*Von Middendorff, A. (1859). "Die Isepiptesen Rußlands".Mem. Acad. Sci. St Petersbourg VI, Ser. Tome* ***8****: 1–143.*

https://en.wikipedia.org/wiki/Domestic_pigeon - cite_ref-Levi_3-0 *Levi, Wendell (1977). The Pigeon. Sumter, S.C.: Levi Publishing Co, Inc.* ISBN 0-85390-013-2.

Crow

Bird Life International *(2012).* "Corvus splendens". IUCN Red List of Threatened Species. *Version 2013.2.*International Union for Conservation of Nature. Retrieved26 November 2013.

https://en.wikipedia.org/wiki/House_crow - cite_ref-2Rasmussen, P.C. and J.C. Anderton (2005) Birds of South Asia: The Ripley Guide. Smithsonian Institution and Lynx Edicions. Vol 2. p.598

https://en.wikipedia.org/wiki/House_crow - cite_ref-4*Cooper, John E. (1996).* "Health studies on the Indian house crow (Corvus splendens)". *Avian Pathology* **25** *(2): 381–386.* doi:10.1080/03079459608419148.

https://en.wikipedia.org/wiki/House_crow - cite_ref-5Ottens, G. (2003). Background and development of the Dutch population of House Crows *Corvus splendens*. *Limosa*, 76(2):69-74

https://en.wikipedia.org/wiki/House_crow - cite_ref-6Pranty, W. (2004). Florida's exotic avifauna: a preliminary checklist. - Birding, August 2004: 362:372.

https://en.wikipedia.org/wiki/House_crow - cite_ref-7Brook, B.W., Sodhi, N.S., Soh, M.C.K. and Lim, H.C. (2003) Abundance and projected control of invasive house crows in Singapore. *Journal of Wildlife Management*, 67(4):808-817

Bat

https://en.wikipedia.org/wiki/Indian_flying_fox - cite_ref-2Old World Fruit Bats (2012): Pteropodidae – Indian Flying Fox (pteropus Giganteus): Species Accounts – Foxes, Trees, Sites, and Pakistan – JRank Articles. Animals.jrank.org. Retrieved on 2012-12-29.

Molur, S., Srinivasulu, C., Bates, P. and Francis, C. (2008).*Pteropus giganteus*. In: IUCN 2012. IUCN Red List of Threatened Species. Version 2012.2.

National Geographic, (2007). "Deadly Contact," David Quammen, pp. 78-105

Monkey

Brandon-Jones, D. (2004). A taxonomic revision of the langurs and leaf monkeys (Primates: Colobinae) of South Asia. *Zoos' Print Journal*, 19 (8): 1552–1594.

Brandon-Jones, D., Eudey, A. A., Geissmann, T., Groves, C. P., Melnick, D. J., Morales, J. C., Shekelle, M. and Stewart, C.B. (2004). Asian Primate Classification *(PDF). 25, No. 1. International Journal of Primatology. pp. 97–164.* Retrieved 5 January 2016.

Burnie D. and Wilson D.E. (2005). Animal: The Definitive Visual Guide to the World's Wildlife. DK Adult (2005), ISBN 0-7894-7764-5.

Groves C. (2001). Primate taxonomy. Smithsonian Institution Press. ISBN 1-56098-872-X

Groves, C.P. *(2005).* Mammal Species of the World: A Taxonomic and Geographic Reference *(3rd ed.). Baltimore: Johns Hopkins University Press. p. 163.* OCLC 62265494. ISBN 0-801- 88221-4.

Groves, C.P. and Molur, S. (2008). "Semnopithecus ajax". IUCN Red List of Threatened Species. *Version 2010.4.* International Union for Conservation of Nature.

https://en.wikipedia.org/wiki/Bonnet_macaque - cite_ref-fooden_3-0Fooden, Jack (1988). Taxonomy and Evolution of the Sinica Group of Macaques: 6. Interspecific Comparisons and Synthesis. *Fieldiana Zoology* (New Series) 45.

https://en.wikipedia.org/wiki/Gray_langur - cite_ref-7*Roonwal, M.L. (1984). Tail form and carriage in Asian and other primates, and their behavioral and evolutionary significance. Current primate research. Jodhpur University Press. pp. 93– 151.*

https://en.wikipedia.org/wiki/Rhesus_macaque - cite_ref-3*Ciani, A. C. (1986).* Inter troop Agonistic Behavior of a Feral Rhesus Macaque Troop in Ranging in Town and Forest Areas in India *(PDF). Aggressive Behavior, 12: 433–439.* Retrieved 5 January 2016.

https://en.wikipedia.org/wiki/Rhesus_macaque - cite_ref-4*Rishi, Kumar; Anindya, Sinha and Sindhu, Radhakrishna (2013). "Comparative Demography of Two Commensal Macaques in India: Implications for Population Status and Conservation". Folia Primatologica, **84**: 384–393.*doi:10.1159/000351935.

https://en.wikipedia.org/wiki/Rhesus_macaque - cite_ref-5*Kumar, R. Radhakrishna and Sinha, A. (2011).* "Of Least Concern? Range Extension by Rhesus Macaques (Macaca mulatta) Threatens Long-Term Survival of Bonnet Macaques (M. radiata) in Peninsular India". *International Journal of Primatology,* **32** *(4): 945– 959.* doi:10.1007/s10764-011-9514-y. Retrieved 5 January 2016.

https://en.wikipedia.org/wiki/Rhesus_macaque - cite_ref-6Jaeger, E. (1972). A source-book of biological names and terms. Springfield, Illinois: Charles C. Thomas.

Kumar, A., Yongzu, Z. and Molur, S. (2008). "Semnopithecus schistaceus". IUCN Red List of Threatened Species. *Version 2010.4.* International Union for Conservation of Nature.

Osterholz, M., Walter, L. and Roos, C. (2008). "Phylogenetic position of the langur genera Semnopithecus and Trachypithecus among Asian colobines, and genus affiliations of their species groups". *BMC Evolutionary Biology, 8: 58.* doi:10.1186/1471-2148-8-58 . PMC 2268674.PMID 18298809.

Shilai, JXWYM (1991). Taxonomic revision and distribution of subspecies of Rhesus Monkey (*Macaca mulatta*) in China. *Zoological Research,* 1991-03 Abstract.

Singh, M., Kumar, A. and Molur, S. (2008). *Macaca radiata*. In: IUCN 2008. IUCN Red List of Threatened Species. Retrieved 24 Nov 2008.

Timmins, R. J., Richardson, M., Chhangani, A. and Yongcheng, L. (2008). "Macaca mulatta". IUCN Red List of Threatened Species. *Version 2010.4.* International Union for Conservation of Nature. Retrieved 5 January 2016.

Wild Boar

Affenberg, W. (1981), The Behavioral Ecology of the Komodo Monitor, University Press of Florida, pp. 248, ISBN 081300621X.

Baskin, L. and Danell, K. (2003). Ecology of Ungulates: A Handbook of Species in Eastern Europe and Northern and Central Asia, Springer Science and Business Media, pp. 15-38, ISBN 3- 540-43804-1

Chen, K. (2007)."Genetic Resources, Genome Mapping and Evolutionary Genomics of the Pig (*Sus scrofa*)". *Int J Biol Sci,* 3(3):153-165. doi:10.7150/ijbs.3.153. Available from http://www. Ijbs.com/v03p0153.htm.

Groves, C. (2008). Current views on the taxonomy and zoogeography of the genus Sus. pp. 15– 29 in Albarella, U., Dobney, K, Ervynck, A. and Rowley-Conwy, P. Eds. (2008). Pigs and Humans: 10,000 Years of Interaction. Oxford University Press. ISBN 978-0-19-920704-6

Groves, C. P.(1993). The Eurasian Suids Sus and Babyrousa. In Oliver, W. L. R., ed., Pigs, Peccaries, and Hippos – 1993 Status Survey and Conservation Action Plan, 107-108. IUCN/SSC Pigs and Peccaries Specialist Group, ISBN 2-8317-0141-4.

Hemmer, H. (1990), Domestication: The Decline of Environmental Appreciation, Cambridge University Press, pp. 55-59,ISBN 0-521-34178-7

Heptner, V. G., Nasimovich, A. A., Bannikov, A. G. and Hoffman, R. S. (1988). Mammals of the Soviet Union, Volume I, Washington, D.C.: Smithsonian Institution Libraries and National Science Foundation, pp. 19-82.

https://en.wikipedia.org/wiki/Wild_boar - cite_ref-Kingdon_16-0Kingdon, J. (1997). The Kingdon Guide to African Mammals. p. 329. Academic Press Limited. ISBN 0-12-408355-2

https://en.wikipedia.org/wiki/Wild_boar - cite_ref-ruvinsky_13-0 Ruvinsky, A. (2011). "Systematics and evolution of the pig". In: Ruvinsky A, Rothschild MF (eds), The Genetics of the Pig. 2nd ed. CAB International, Oxon. pp. 1-13. ISBN 978-1- 84593-756-0.

Kurtén, Björn (1968). Pleistocene mammals of Europe. Weidenfeld and Nicolson. pp. 153-155.

Oliver, W. and Leus, K. (2008). *Sus scrofa*. In: IUCN 2008. IUCN Red List of Threatened Species. Retrieved 6 March 2013. Database entry includes a brief justification of why this species is of least concern.

Oliver, W. L. R. (1993). The Eurasian Wild Pig (*Sus scrofa*). In Oliver, W. L. R., ed., Pigs, Peccaries, and Hippos – 1993 Status Survey and Conservation Action Plan, 112-121. IUCN/SSC Pigs and Peccaries Specialist Group, ISBN 2-8317-0141-4

Wozencraft, W.C. (2005). "Order Carnivora". *In Wilson, D.E.; Reeder, D.M.* Mammal Species of the World: A Taxonomic and Geographic Reference *(3rd ed.). Johns Hopkins University Press. pp. 532–628.* ISBN 978-0-8018-8221-0.OCLC 62265494.

Blue Bull

D.M. Leslie (2008). "Boselaphus tragocamelus (Artiodactyla: Bovidae)". *Mammalian Species,* ***813****: Number 813: pp. 1–16.*doi:10.1644/813.1.

Hassanin, A. and Douzery, E. J. P. (1999). "Evolutionary affinities of the enigmatic saola (Pseudoryx nghetinhensis) in the context of the molecular phylogeny of Bovidae".Proceedings of the Royal Society B: Biological Sciences, **266** *(1422): 893–900.*doi:10.1098/rspb.1999.0720.

https://en.wikipedia.org/wiki/Nilgai - cite_ref-allard_13-0 *Allard, M.W., Miyamoto, M.M., Jarecki, L., Kraus, F. and Tennant, M.R. (1992). "DNA systematics and evolution of the artiodactyl family Bovidae". Proceedings of the National Academy of Sciences of the United States of America,* ***89*** *(9): 3972–6.* PMID 1570322.

https://en.wikipedia.org/wiki/Nilgai - cite_ref-brain_15-0*Brain, C.K. (1981). The Hunters or the Hunted?: An Introduction to African Cave Taphonomy. Chicago, USA: University of Chicago Press. p. 167.* ISBN 978-0-226-07089-6.

https://en.wikipedia.org/wiki/Nilgai - cite_ref-MSW3_7-0*Wilson, D.E. and Reeder, D.M. (2005).* Mammal Species of the World: A Taxonomic and Geographic Reference *(3rd ed.). Johns Hopkins University Press.* ISBN 978-0-8018-8221-0. OCLC 62265494.

Mallon, D.P. (2008). *Boselaphus tragocamelus.* In:IUCN 2008. IUCN Red List of Threatened Species. Retrieved 29 March 2009. Database entry includes a brief justification of why this species is of least concern.

Padhi, S., Panigrahi, G.K. and Panda, S. (2004). The Wild Animals of India. Delhi: Biotech Books. pp. 26–7. ISBN 81-7622-106-6.

Elephant

Choudhury, A., Lahiri Choudhury, D. K., Desai, A., Duckworth, J. W., Easa, P. S., Johnsingh, A. J. T., Fernando, P., Hedges, S., Gunawardena, M., Kurt, F., Karanth, U., Lister, A., Menon, V., Riddle, H., Rübel, A., Wikramanayake, E. (2008). "Elephas maximus". IUCN Red List of Threatened Species. *Version 2012.2.*International Union for Conservation of Nature.

https://en.wikipedia.org/wiki/Indian_elephant - cite_ref-10Samansiri, K. A. P. and Weerakoon, D. K. (2007). Feeding Behaviour of Asian Elephants in the Northwestern Region of Sri Lanka. Gajah: *Journal of the IUCN/SSC Asian Elephant Specialist Group,* Number 2: 27–34

https://en.wikipedia.org/wiki/Indian_elephant - cite_ref-11Sukumar, R. (1990). Ecology of the Asian Elephant in southern India. II. Feeding habits and crop raiding patterns. *Journal of Tropical Ecology,* 6: 33–53.

https://en.wikipedia.org/wiki/Indian_elephant - cite_ref-12Pradhan, N.M.B., Wegge, P., Moe, S.R. and Shrestha, A.K. (2008). Feeding ecology of two endangered sympatric megaherbivores: Asian elephant Elephas maximus and greater one-horned rhinoceros Rhinoceros unicornis in lowland Nepal. *Wildlife Biology,* 14: 147–154.

https://en.wikipedia.org/wiki/Indian_elephant - cite_ref-13Borah, J., Deka, K. (2008). Nutritional Evaluation of Forage Preferred by Wild Elephants in the Rani Range Forest, Assam, India. Gajah: Journal of the IUCN/SSC *Asian Elephant Specialist Group,* 28: 41–43.

https://en.wikipedia.org/wiki/Indian_elephant - cite_ref-2Shoshani, J. and Eisenberg, J. F. (1982). *Elephas maximus. Mammalian Species,* 182: 1–8.

https://en.wikipedia.org/wiki/Indian_elephant - cite_ref-4Pillai, N.G. (1941). On the height and age of an elephant. *Journal of the Bombay Natural History Society,* 42: 927–928.

https://en.wikipedia.org/wiki/Indian_elephant - cite_ref-5Furaha tenVelde, P. (1997). The wild elephants of the Royal Bardia National Park, Nepal. Gajah: *Journal of the IUCN/SSC Asian Elephant Specialist Group,* 17: 41–44.

https://en.wikipedia.org/wiki/Indian_elephant - cite_ref-6*Ben S. Roesch.(2006).* "Living Stegodont or Genetic Freak?". *Archived from the original on 8 November 2006*. Retrieved2008-06-18.

https://en.wikipedia.org/wiki/Indian_elephant - cite_ref-9Bhatta, S. R. (2006). Efforts to conserve the Asian elephant in Nepal. Gajah: Journal of the IUCN/SSC Asian Elephant Specialist Group 25: 87–89.

https://en.wikipedia.org/wiki/Indian_elephant - cite_ref-Shoshani_3-0Shoshani, J. (2006). Taxonomy, Classification, and Evolution of Elephants In: Fowler, M. E., Mikota, S. K. (eds.) Biology, medicine, and surgery of elephants. Wiley-Blackwell. ISBN 0- 8138-0676-3. pp. 3–14.

Sukumar, R. (1993). The Asian Elephant: Ecology and Management Second edition. Cambridge University Press. ISBN 0-521-43758-X.

Index

O

P